# Reconceptualizing the Nature of Science for Science Education

# Contemporary Trends and Issues in Science Education

## VOLUME 43

### SERIES EDITOR
Dana Zeidler, *University of South Florida, Tampa, USA*

### FOUNDING EDITOR
Ken Tobin, *City University of New York, USA*

### EDITORIAL BOARD
Fouad Abd El Khalick, *University of Illinois at Urbana-Champaign, USA*
Marrisa Rollnick, *University of the Witwatersrand, Johannesburg, South Africa*
Svein Sjøberg, *University of Oslo, Norway*
David Treagust, *Curtin University of Technology, Perth, Australia*
Larry Yore, *University of Victoria, British Columbia, Canada*
HsingChi von Bergmann, *University of Calgary, Canada*
Troy D. Sadler, *University of Missouri, Columbia, USA*
Michael P. Clough, *Iowa State University, Ames, IA, USA*

### SCOPE
The book series Contemporary Trends and Issues in Science Education provides a forum for innovative trends and issues connected to science education. Scholarship that focuses on advancing new visions, understanding, and is at the forefront of the field is found in this series. Accordingly, authoritative works based on empirical research and writings from disciplines external to science education, including historical, philosophical, psychological and sociological traditions, are represented here.

More information about this series at http://www.springer.com/series/6512

Sibel Erduran • Zoubeida R. Dagher

# Reconceptualizing the Nature of Science for Science Education

Scientific Knowledge, Practices and Other Family Categories

Springer

Sibel Erduran
Department of Education
 and Professional Studies
University of Limerick
Limerick, Ireland

Zoubeida R. Dagher
School of Education
University of Delaware
Newark, DE, USA

ISSN 1878-0482     ISSN 1878-0784 (electronic)
ISBN 978-94-017-9056-7     ISBN 978-94-017-9057-4 (eBook)
DOI 10.1007/978-94-017-9057-4
Springer Dordrecht Heidelberg New York London

Library of Congress Control Number: 2014947967

© Springer Netherlands 2014
This work is subject to copyright. All rights are reserved by the Publisher, whether the whole or part of the material is concerned, specifically the rights of translation, reprinting, reuse of illustrations, recitation, broadcasting, reproduction on microfilms or in any other physical way, and transmission or information storage and retrieval, electronic adaptation, computer software, or by similar or dissimilar methodology now known or hereafter developed. Exempted from this legal reservation are brief excerpts in connection with reviews or scholarly analysis or material supplied specifically for the purpose of being entered and executed on a computer system, for exclusive use by the purchaser of the work. Duplication of this publication or parts thereof is permitted only under the provisions of the Copyright Law of the Publisher's location, in its current version, and permission for use must always be obtained from Springer. Permissions for use may be obtained through RightsLink at the Copyright Clearance Center. Violations are liable to prosecution under the respective Copyright Law.
The use of general descriptive names, registered names, trademarks, service marks, etc. in this publication does not imply, even in the absence of a specific statement, that such names are exempt from the relevant protective laws and regulations and therefore free for general use.
While the advice and information in this book are believed to be true and accurate at the date of publication, neither the authors nor the editors nor the publisher can accept any legal responsibility for any errors or omissions that may be made. The publisher makes no warranty, express or implied, with respect to the material contained herein.

Printed on acid-free paper

Springer is part of Springer Science+Business Media (www.springer.com)

*A genuine and engaging emphasis on ideas is necessary in determining the curriculum. We should be fearless about ideas and openly engage in discussion and debate about what should matter in the subject matter.*

Heidi Hayes Jacobs, *Curriculum 21* (2012)

*I dedicate this book to my family Ayten, Erol and Suna Erduran for their sense of compassion and justice. (SE)*

*I dedicate this book to my parents Nadra and Ramez Dagher, my husband Bahgat Bichara, and children Grace and David, for their unconditional love and support.*

# Foreword

Teaching nature of science has for decades been an accepted goal of science education. The goal, however, has been fraught with controversy. Is the goal to teach *the* nature of science, or is it to teach *about* the nature of science. If the goal is taken as the former of these options, then the question gets raised, "Whose nature of Science?" The question is motivated by the correct perception that the nature of science is itself fraught with controversy. Although there are lots of settled ideas belonging to substantive scientific content that can be taught without controversy, when it comes to teaching nature of science, controversy and conflicting views are the norm. One proposed solution to this apparent lack of content to teach is to distil from the vast literature on the nature of science propositions on which there is a consensus of agreement. Such propositions as the following have been proposed: scientific knowledge is tentative; science has an empirical basis; scientific knowledge is not simply discovered but there is a sense in which it is also constructed by scientists. Problem is that such statements are little more than bromides that poorly serve curriculum development. What guidance is provided by such statements on, for example, the developmental progression of nature of science topics from primary school to high school?

The other proposed solution is to adopt the latter goal, that is, to teach about the nature of science. There is much content available from history, philosophy, and sociology of science. However, science educators are not seeking and do not want a course in the history, philosophy, or sociology of science. As valuable as such courses might be in other contexts, they are not what science teachers desire as part of science courses. So, we seem stuck, and that is the condition in which many have perceived science education to be when it comes to teaching nature of science. Is there a way forward that can preserve the conviction that teaching nature of science is important to science education by supporting it with a sound theoretical foundation?

It is into this contested zone that Sibel Erduran and Zoubeida Dagher have leapt with a proposed solution. They draw upon the family resemblance framework of Irzik and Nola (2013) and extend that framework to create a scheme for teaching nature of science that is broader than typically found. The scheme addresses not only the usual cognitive and epistemic aspects of science but also social-institutional aspects that often are not considered in school science teaching, such as the political and financial dimensions of science. Their rationale for adopting this broader approach is that otherwise undue attention is paid to the factors that influence the development and validation of scientific claims. This rationale is one that I heartily endorse, because it characterizes science both as a cognitive-epistemic system and as a social-institutional system. As Erduran and Dagher convincingly demonstrate with a copious use of examples throughout the volume, this family resemblance approach can serve as a powerful and useful reminder of nature of science considerations that need to arise in specific contexts and of what is missing.

The volume is structured so as to articulate their theoretical stance is a systematic manner. Chapter 1 provides a motivation for the work and their rationale for adopting and revising the family resemblance framework of Irzik and Nola. The middle six chapters explain and illustrate through examples and functional visualizations the main features of the framework. These six chapters bear the theoretical weight of the work and the authors provide persuasive grounds for accepting what already in Chap. 1 had been presented as a framework with intuitive appeal.

The capstone chapter, Chap. 8, brings the arguments of the previous chapters into a coherent whole, by demonstrating how the family resemblance framework is capable of unifying nature of science content for a science curriculum extending from the primary to secondary grades. This accomplishment can be claimed by no other work that is familiar to me. The visualizations that had been introduced in the previous chapters are brought forward once again in Chap. 8, this time as "Generative Images of Science". The metaphor is compelling. The authors are able to show the potential of each image to help in the articulation of the aspect of science it depicts for use in primary school to high school science teaching. The images direct attention to all of the different components of the nature of science and demand answers to the question: How can each of these components be realized in the science we are here attempting to teach? Thus, the articulation of the family resemblance framework occurs both vertically (across all grade levels) and horizontally (within science topics). Thus, across 12 years of schooling, students will have seen each aspect of the nature of science arise in each of their grades across many topics. As I understand it, their approach relies not on the promulgation of true propositions about nature of science (although it is not antagonistic to such propositions that do exist), but on intelligent classroom discussions about issues regarding nature of science as they arise in learning substantive science content.

Finally, perhaps we have the theoretical formulation capable to ground an actual nature of science curriculum that manifests developmental progression over the 12 years of schooling and that fully integrates nature of science with substantive science content. Such a formulation is surely needed, because, as the authors show, even our most up-to-date curriculum policy documents fail to present an articulate vision of how nature of science is to be taught so as to justify holding it up as a goal of science education for all of these previous decades. I commend Sibel Erduran and Zoubeida Dagher for their daring and highly intelligible statement on this most important of science education topics.

Department of Educational Policy Studies,                         Stephen P. Norris
University of Alberta, Edmonton, AB, Canada
February 7, 2014

# Preface

While we share a longtime interest in the philosophy of science in science education, the first thought of writing this book was triggered by a symposium during the 2010 annual meeting of *NARST: A Worldwide Organization for Improving Science Education through Research on Teaching and Learning* in Philadelphia, USA. Among the sentiments strongly voiced in that symposium was an admonition to stop asking what science is, for, it was argued the content of what needs to be taught is already known. What is needed instead, the argument continued, was better ways of incorporating nature of science in science teaching. We disagreed. From our perspective, the question of 'what' nature of science to include in science education has been addressed by some researchers but is far from being settled. We became seriously concerned that a critical attitude about nature of science content has been turned off. "Nature of science" seemed beyond questioning. But nothing is beyond questioning especially in science which is an ever developing enterprise. We were also concerned about the promulgation of overly generic accounts of nature of science that did not attend to domain specificity of science disciplines. The idea of this book was thus born, with the aim of fostering a critical and constructive debate about how to reconceptualize nature of science for science education.

Our primary goal was to synthesize new ideas on how nature of science can be considered in science education so that learners of science can be inspired by the awe and wonder of the many faces of science and learn to think scientifically. In the spirit of scientific reasoning, we wanted to have an evidence-based approach in characterizing the nature of science. This notion has led us to the vast philosophy of science literature focused on the various science disciplines. We immersed ourselves in this literature which enriched our understanding of some contemporary debates on the nature of science. As science educators, we were not interested in philosophy of science for philosophy's sake, but rather we used philosophy of science to achieve conceptual clarity about what we want science lessons to include about nature of science. The experience has taught us that it is vital for science educators to be mindful of first-hand accounts in the philosophy of science and other relevant foundational disciplines such as history, anthropology and psychology of science.

Among the perspectives that we considered, a germ of a fruitful idea in the Family Resemblance Approach proposed by philosophers of science Gurol Irzik and Robert Nola seemed appropriate for our inquiry. We were particularly inspired by their plenary lecture at the *International History and Philosophy in Science Teaching [IHPST]* Conference held in Thessaloniki, Greece, in 2011, in which they had expanded their earlier published account. The Family Resemblance Approach provided us with a unifying yet flexible framework for promoting a relatively broad and inclusive account of nature of science for science education, one that acknowledges common features while at the same time accommodating disciplinary particularities.

We have not only gone beyond Irzik and Nola's depictions of nature of science (for instance, by both expanding their framework and adding more categories to it) but also transformed the ideas into pedagogically sound opportunities. One of the key avenues of transformation was the introduction of visual representations on the various 'family' categories to facilitate not only the communication of some rather deep philosophical issues but also to provide practical toolkits for educators and researchers. We have shared themes from this book with researchers, teacher educators and teachers at professional conferences such as NARST, ESERA, IHPST, ECER, Improving Middle School Science Instruction Using Cognitive Science, Washington, DC; as well in plenary talks at the Annual Science and Math Educators Conference at the American University of Beirut, Lebanon; WCNSTE, Poland; IOSTE Eurasia Regional Conference, Turkey; Frontiers in Mathematics and Science Education Research Conference at Eastern Mediterranean University, Cyprus; and the European Conference on Research in Chemistry Education, Finland.

In working on this book project, we realized that we share similar values about respecting diversity and inclusion of ideas, learners and strategies in educational processes. As individuals whose childhoods were spent in areas of the world torn by political and armed conflict (Erduran in Cyprus and Dagher in Lebanon) we also possess propensity to reconcile different points of view, to move beyond stagnation and to propose constructive dialogue for improving education. Our appreciation of complexity and love for holistic accounts were great motivators although they taxed our time and brains. We took on the challenge of bringing together conventionally disparate ideas, for instance, philosophical reflection and practical teaching concerns. We believe that it is our professional imperative to embrace such challenges and to debate issues openly. Inevitably, work of this kind will be limited by nature. For this reason, we invite colleagues and future researchers to extend our work in order to contribute further to the study of nature of science in science education.

There are many scholars who have shaped our orientation to the field throughout our careers. In addition to our doctoral research mentors, Richard Duschl and George W. Cossman, we acknowledge the writings of Joseph Schwab, Douglas Roberts, Stephen Norris and Michael Matthews that have provided motivation for pursuing scholarship in this field. We continue to draw inspiration from the contributions of other colleagues, too numerous to name here, who are engaged in tireless efforts to develop science education theory and practice from diverse foundational perspectives.

# Preface

We are grateful to the feedback from two anonymous reviewers. We particularly wish to thank Gurol Irzik who read an earlier draft and provided valuable comments. We are also grateful to Stephen Norris, for taking the time to read the manuscript and writing the generous Foreword that he shared with us mere 11 days before his sudden passing. We are extremely saddened by his loss and regret that we did not have an opportunity to thank him in person. Stephen was a remarkably decent human being and a fantastic colleague who will always be remembered for his critical mind, genuine kindness, and great sense of humor. The rigor of his thinking and depth of his knowledge will continue to inspire us.

We wish to thank Megan F. Byrne for her dedication in proofreading an earlier version of the book manuscript. Additional feedback from her perspective as an elementary teacher education major at the University of Delaware was useful in helping us improve the clarity of the text. Zoubeida Dagher acknowledges the support of the Center for Science, Ethics, and Public Policy at the University of Delaware at different stages of writing this book. We would also like to extend our thanks to Bernadette Ohmer at Springer for useful, timely and supportive interactions throughout the manuscript preparation process. The arguments were co-developed in the course of conversations and writing sessions facilitated by an arsenal of communication and file sharing tools such as Skype, GoogleDrive, Dropbox, and e-mail. Our collaboration has benefited greatly from these marvels of contemporary technology as well as a good dose of mutual sense of humor.

Finally, we are grateful for the support of our families and friends.

| | |
|---|---|
| Limerick, Ireland | Sibel Erduran |
| Newark, DE, USA | Zoubeida R. Dagher |

# Contents

**1 Reconceptualizing Nature of Science for Science Education** .............. 1
   1.1  Introduction .................................................................................. 1
   1.2  Brief History of NOS in Science Education Research ................. 3
   1.3  NOS in Curricular Context ........................................................... 9
   1.4  Key Contributions of the Book ..................................................... 13
   References ............................................................................................. 16

**2 Family Resemblance Approach to Characterizing Science** ................ 19
   2.1  Introduction .................................................................................. 19
   2.2  Justifying the Family Resemblance Approach ............................. 24
   2.3  Extending the Family Resemblance Approach ............................ 27
   2.4  The FRA as a Holistic Model ....................................................... 28
   2.5  The Relationship of FRA to Research Traditions
        and Policy in Science Education .................................................. 31
   2.6  Potential Challenges in Applying the FRA
        in Science Education .................................................................... 33
   2.7  The Layout of the Book ............................................................... 36
   References ............................................................................................. 38

**3 Aims and Values of Science** ................................................................... 41
   3.1  Introduction .................................................................................. 41
   3.2  What Are Aims and Values of Science? ....................................... 43
   3.3  Generating a Framework on Scientific Aims
        and Values for Science Education ................................................ 48
        3.3.1  What Are Aims and Values in Science? .......................... 48
        3.3.2  How Do Aims and Values Function in Science? ............. 49
   3.4  Educational Applications ............................................................. 50
   3.5  Fostering Scientific Aims and Values in Science Education ........ 54
   3.6  Conclusions .................................................................................. 57
   References ............................................................................................. 64

## 4 Scientific Practices .................................................. 67
- 4.1 Introduction .................................................. 67
- 4.2 Differentiating Scientific Practices, Processes and Activities ........... 69
- 4.3 Examples of Scientific Practices: Classification, Observation and Experimentation .................................................. 71
- 4.4 A Proposed Heuristic of Scientific Practices ........... 80
- 4.5 Application of the Benzene Ring Heuristic ........... 83
- 4.6 Conclusions and Discussion ........... 85
- References .................................................. 86

## 5 Methods and Methodological Rules ........... 91
- 5.1 Introduction .................................................. 91
- 5.2 Beyond the "Scientific Method" ........... 92
- 5.3 Scientific Methods and Methodological Rules ........... 96
- 5.4 Methodological Rules as Evolving Entities ........... 104
- 5.5 Educational Implications ........... 105
- 5.6 Conclusions ........... 111
- References .................................................. 111

## 6 Scientific Knowledge ........... 113
- 6.1 Introduction .................................................. 113
- 6.2 Classification of Scientific Knowledge Forms ........... 116
- 6.3 Domain-Specificity of Scientific Knowledge ........... 120
- 6.4 Evaluation of Scientific Knowledge ........... 125
- 6.5 Explanatory Dimension of Scientific Knowledge ........... 126
- 6.6 Educational Applications ........... 131
- References .................................................. 132

## 7 Science as a Social-Institutional System ........... 137
- 7.1 Introduction .................................................. 137
  - 7.1.1 Professional Activities ........... 139
  - 7.1.2 Scientific Ethos ........... 140
  - 7.1.3 Social Certification and Dissemination ........... 141
  - 7.1.4 Social Values of Science ........... 142
- 7.2 Elaborating on Science as a Social-Institutional System ........... 142
  - 7.2.1 Social Organisations and Interactions ........... 145
  - 7.2.2 Political Power Structures ........... 146
  - 7.2.3 Financial Systems ........... 148
- 7.3 Educational Applications ........... 150
  - 7.3.1 Teaching and Learning of Science as a Social-Institutional System ........... 150
  - 7.3.2 Curricular Policy ........... 156
- 7.4 Conclusions ........... 160
- References .................................................. 161

| | | |
|---|---|---|
| **8** | **Towards Generative Images of Science in Science Education** | 163 |
| | 8.1 Introduction | 163 |
| | 8.2 Educational Applications of FRA and GIS | 166 |
| |     8.2.1 Vertical Articulation | 167 |
| |     8.2.2 Horizontal Articulation | 171 |
| | 8.3 FRA, GIS and Curriculum Policy Documents | 174 |
| |     8.3.1 Example 1: HS-LS3 Heredity: Inheritance and Variation of Traits | 174 |
| |     8.3.2 Example 2: From Molecules to Organisms – Structures and Processes | 176 |
| | 8.4 Potential limitations of the FRA and GIS | 176 |
| | 8.5 Recommendations | 181 |
| |     8.5.1 Teaching | 181 |
| |     8.5.2 Teacher Education | 181 |
| |     8.5.3 Curriculum and Assessment | 183 |
| | 8.6 Contributions to Research and Practice in Science Education | 183 |
| | 8.7 Conclusions | 185 |
| | References | 186 |
| **Authors Biographies** | | 189 |

# Chapter 1
# Reconceptualizing Nature of Science for Science Education

The chapter sets out the agenda for the entire book. The primary aim is to illustrate how "Nature of Science" (NOS) can be conceptualized and subsequently applied in science education research, policy and practice. Considering the vast amount of research literature in science education on NOS, the intention is to highlight some of the recent debates on the topic and provide a rationale for a new direction in this area. The contribution to the NOS debate is made by appealing to the theoretical grounding of arguments in science education on foundational fields like philosophy of science to ensure consistency with contemporary meta-accounts of science. In other words, an evidence-based theoretical rationale is followed to illustrate what is meant by 'science'. The implications of various investigations into different aspects of science (e.g. epistemic, cognitive and social aspects) are numerous for curriculum content, instructional approaches and learning outcomes. Even though the coverage is theoretical in nature, plenty of concrete examples are used to illustrate how the ideas might translate to the level of the classroom so that they are applicable and relevant for science teachers and learners. Once the theoretical rationale for a new approach is built and unpacked, empirical validation of these ideas may follow, including the testing for the effectiveness of some of the proposed strategies.

## 1.1 Introduction

Two fundamental questions about science are relevant for science educators: (a) what is the nature of science? and (b) what ideas about nature of science should be taught and learned? In order to address these questions in our own professional trajectories as science education researchers, we have often resorted to inquiries into meta-perspectives on science in order to understand what counts as science, what makes science 'science' and what thus should be the content of science education. We were often drawn to meta-perspectives from philosophy of science in particular because upon exposure to this area of scholarship, we recognized that understanding science requires not just understanding of scientific knowledge and processes, but also understanding how we get to understand what science is. In order to be analytical about the nature of science, then, we recognized that appeal to a higher level of analysis and justification of science from a philosophical perspective was essential. We recognized that our notions of 'science' will be limited if they are solely based on what we learned through school science ourselves, especially if we have had limited formal exposure to the culture of working scientists or have not done any authentic scientific inquiry ourselves. Similarly, educational accounts of science present a particular version of science that has already been transformed for teaching and learning purposes. In our exposure to philosophy of science, we have developed and continue to develop an appreciation of the fact that science is a complex endeavor and that despite an agreed set of knowledge and processes that science encapsulates, understanding what is science is an ongoing agenda. In contrast, school science persists in projecting to teachers and students a rather simplistic, narrow and unproblematic account of science. For example, school science rarely coordinates the epistemic, cognitive and social dimensions of science so that learners develop a balanced understanding of what is meant by science in a holistic sense.

How then can we, as science educators, settle on what of science should be included in science education? How can reflective accounts about science be captured in science education? In our pursuits of these questions within science education research literature, we have found several sources that made significant attempts to address these questions. For instance, the rich body of scholarship in the applications of history, philosophy and sociology of science has certainly provided a great deal of perspective on science in science education research and practice (e.g. Matthews, 1994, 2014). The substantial amount of empirical research on the so-called "nature of science" (NOS) (e.g. Abd-El-Khalick, 2012; Abd-El-Khalick & Lederman, 2000; Ackerson & Donnelly, 2008; Lederman, 1992) has further provided an avenue for science educators to engage in these key questions. Hence, one could ask why this book then? What new perspectives can be synthesized from philosophy of science on aspects of science to further the agenda of science education research and practice? In order to tackle such questions, a context for NOS in science education is needed. There is plethora of work on the characterization of science, including the nature of scientific knowledge and scientific inquiry. Our intention in this book is not to review any of these bodies of work in great detail as there is plenty of literature that has already accomplished this very goal (e.g. Abd-El-Khalick & Lederman, 2000; Duschl & Grandy, 2013; Lederman, 1992, 2007; McComas, Clough, & Almazroa, 1998). Our objective is to develop a new direction to build on the content, rationale and application of NOS in science education research, practice and policy.

Our pursuit of the characterization of NOS is guided by a set of principles that include the following: respect for diversity and inclusion; care for motivation and affective dimensions of learning; and social justice in making science and scientific reasoning accessible. The overall goals is the empowerment of students such that their interests and understandings of science can be fostered and nurtured with a multitude of perspectives on science. In re-conceptualizing NOS for science education, inclusion and diversity thus are important. Catering to the needs and interests of a diverse group of learners is not a luxury, but a necessity given the trends in globalization. Documents such as the *Index for Inclusion* (Booth & Ainscow, 2000) illustrate some of the key imperatives for engaging all students in learning processes. The *Index for Inclusion*, which was widely distributed in England and Wales, is a set of materials developed in England to help schools reduce barriers to learning and participation as well as valuing *all* pupils equally.

The key messages of this and similar global documents and initiatives are that learners should be provided with opportunities to participate in effective learning, and any barriers to inclusion of diversity in education should be removed to ensure that there is equity in representation. The ideas presented in the book are necessarily broad and inclusive providing more of a likelihood of participation and access to understanding science. The overall case for the reconceptualization of nature of science in science education rests on coordinating the epistemic, cognitive and social aspects of science for the purposes of supporting a more inclusive portrayal of science in science teaching and learning. A skeptic of our approach might argue otherwise. An opposing argument could be that the very fact that the conventional characterizations of NOS are extended, making the depiction of science more complex, we are only making it accessible to only a select few because curriculum content would need to be far more complex and cognitively demanding to learners. As each chapter and particularly Chap. 8 will detail, the argument is not about the addition of excessive amount of content in the curriculum. Rather, the recommendation is to use the content of the curriculum in more holistic and creative ways in order to present a more comprehensive and balanced account of science to learners. NOS conceptualized in this fashion is inherently inclusive of the many faces of science that are more likely to be motivating to a wider range of students. For instance, some students might be more drawn to the epistemic practices of science while others are very much interested in the socio-political dimensions of the scientific enterprise. By extending the characterization of NOS, we are not only attempting to be faithful to an authentic account of science but also are potentially promoting the participation and engagement of more students in science. Eventually, the merit of the proposed approach will rest on the empirical testing and validation of the proposed strategies. The book thus provides science education researchers with a new territory for innovation and investigation.

## 1.2 Brief History of NOS in Science Education Research

Nature of science (NOS) has become a predominant area of research in science education in the past few decades (Abd-El-Khalick, Bell, & Lederman, 1998; Allchin, 2013; Alters, 1997; Eflin, Glennan, & Reisch, 1999; Lederman, 1992; McComas et al., 1998; McComas, & Olson, 1998; Rubba, & Anderson, 1978; Smith,

Lederman, Bell, McComas, & Clough, 1997). Driver, Leach, Millar, and Scott (1996) have highlighted five potential benefits of students' learning of the nature of science, namely that understanding of the nature of science helps students to (a) understand the process of science, (b) make informed decisions on socio-scientific issues, (c) appreciate science as a pivotal element of contemporary culture, (d) be more aware of the norms of the scientific community, and (e) learn science content with more depth.

Definitions of the nature of scientific knowledge presented in the science education literature are diverse. The work in the 1960s included seminal pieces by Conant (1961) and Klopfer (1969). According to Klopfer, the processes of scientific inquiry and the developmental nature of knowledge acquisition in science depict the nature of science. Klopfer identifies the understanding of how scientific ideas are developed as one of three important components of scientific literacy. In this view, students must learn how scientific ideas are formulated, tested, and inevitably, revised, and he/she must learn what motivates scientists to engage in this activity. Kimball (1968) developed a model of the nature of science following an extensive review of literature on the nature and philosophy of science. The main statements guiding his model were the following:

(1) The fundamental driving force in science is curiosity concerning the physical universe. It has no connection with outcomes, applications, or uses aside from the generation of new knowledge.
(2) In the search for knowledge, science is process-oriented; it is a dynamic, ongoing activity rather than a static accumulation of information.
(3) In dealing with knowledge as it is developed and manipulated, science aims at ever-increasing comprehensiveness and simplification, emphasizing mathematical language as the most precise and simplest means of stating relationships.
(4) There is no one "scientific method" as often described in school science text-hooks. Rather, there are as many methods of science as there are practitioners.
(5) The methods of science are characterized by a few attributes which are more in the realm of values than techniques. Among these traits of science are dependence upon sense experience, insistence on operational definitions, …., and the evaluation of scientific work in terms of reproducibility and of usefulness in furthering scientific inquiry.
(6) A basic characteristic of science is a faith in the susceptibility of the physical universe to human ordering and understanding.
(7) Science has a unique attribute of openness, both openness of mind, allowing for willingness to change opinion in the face of evidence, and openness of the realm of investigation, unlimited by such factors as religion, politics, or geography.
(8) Tentativeness and uncertainty mark all of science. Nothing is ever completely proven in science, and recognition of this fact is a guiding consideration of the discipline. (Kimball, 1968, pp. 111–112)

Some of the work conducted in the 1970s included that of Showalter (1974) who used the concepts tentative, public, replicable, probabilistic, humanistic, historic, unique, holistic, and empirical to characterize the nature of scientific knowledge. After conducting a review of literature on the nature of scientific knowledge, Rubba and Anderson (1978) consolidated the nine concepts identified by Showalter into a six-factor model called "A Model of the Nature of Scientific Knowledge". The six factors included by Rubba and Anderson are defined as amoral (scientific knowledge itself cannot be judged as morally good or bad), creative (scientific knowledge

## 1.2 Brief History of NOS in Science Education Research

is partially a product of human creativity), developmental (scientific knowledge is tentative), parsimonious (scientific knowledge attempts to achieve simplicity of explanation as opposed to complexity), testable (scientific knowledge is capable of empirical test), and unified (the specialized sciences contribute to an interrelated network of laws, theories, and concepts).

Other researchers such as Cotham and Smith (1981) use the terms 'tentative' and 'revisionary' to describe the nature of scientific theories. The tentative component of this conception highlights the inconclusiveness of all knowledge claims in science. The revisionary component indicates the revision of existing scientific knowledge in response to changing theoretical frameworks. While NOS has been used as terminology in the literature to represent the same facets as scientific knowledge, it is usually presented in a broader context. This broader context includes not only the nature of scientific knowledge, but the nature of the scientific enterprise and the nature of scientists as well (Cooley & Klopfer, 1963).

More contemporary accounts of NOS in the science education research literature have been reviewed by Chang, Chang, and Tseng (2010) who traced the literature between 1990 and 2007. The key proponents during this period in science education (Abd-El-Khalick, 2012; Lederman et al., 2002; McComas, 1998) have outlined a set of statements that characterize what has been referred to as a "consensus view" of the nature of science. The key aspects of this approach are as follows:

1. The Tentative Nature of Scientific Knowledge: Scientific knowledge, although reliable and durable, is never absolute or certain. This knowledge, including facts, theories, and laws, is subject to change.
2. Observation, Inference, and Theoretical Entities in Science: Observations are descriptive statements about natural phenomena that are directly accessible to the senses (or extensions of the senses). By contrast, inferences are statements about phenomena that are not directly accessible to the senses.
3. The Theory-Laden Nature of Scientific Knowledge: Scientific knowledge is theory-laden. Scientists' theoretical and disciplinary commitments, beliefs, prior knowledge, training, experiences, and expectations actually influence their work.
4. The Creative and Imaginative Nature of Scientific Knowledge: Science is empirical. Nonetheless, generating scientific knowledge also involves human imagination and creativity.
5. The Social and Cultural Embeddedness of Scientific Knowledge: Science as a human enterprise is practiced in the context of a larger culture and its practitioners are the product of that culture.
6. Scientific theories and laws: Scientific theories are well-established, highly substantiated, internally consistent systems of explanations. Laws are descriptive statements of relationships among observable phenomena. Theories and laws are different kinds of knowledge and one does not become the other.
7. Myth of The Scientific Method: The myth of the scientific method is regularly manifested in the belief that there is a recipelike stepwise procedure that all scientists follow when they do science. This notion was explicitly debunked (Lederman et al., 2002, pp. 500–502).

The "consensus view" of NOS has led to a major body of empirical studies in science education (Abd-El-Khalick & Lederman, 2000; Ackerson & Donnelly, 2008). While many science educators agree with the key tenets of this definition of NOS, several points of debate have been prevalent in the community. For example, some authors (e.g. Lederman, 2007) have advised that while NOS and scientific inquiry are related, they should be differentiated. The main premise of this argument is that 'inquiry' can be specified as the methods and procedures of science while the NOS concerns more the epistemological features of scientific processes and knowledge.

Grandy and Duschl (2007) have disputed such distinctions on the basis that they "greatly oversimplify the nature of observation and theory and almost entirely ignores the role of models in the conceptual structure of science" (p. 144). Although Lederman (2007) advocates using the phrase "nature of scientific knowledge" (rather than NOS) to avoid the conflation issue, scientific inquiry (especially "scientific methods") has been considered an important aspect of NOS in other researchers' work (e.g. Ryder, Leach, & Driver, 1999). A related set of research studies highlight the epistemological goal of inquiry (e.g. Sandoval, 2005) and epistemological enactment through inquiry (e.g. Ford, 2008).

Further critiques of the consensus view have been leveled by other science educators. Clough (2007) has suggested that the tenets proposed by the consensus view, should be turned from declarative statements into questions that promote discussion about the nature of science. Yacoubian (2012) argued that the consensus views: "(1) lack clarity in terms of how NOS-related ideas could be applied for various ends, (2) portray a distorted image of the substantive content of NOS and the process of its development, and (3) lack a developmental trajectory for how to address NOS at different grade levels." (p. 25) He proposes ways to rectify these issues using a critical thinking CT-NOS framework to create developmentally appropriate NOS lessons.

Recognizing the importance of broadening the scope of nature of science beyond the consensus view, Matthews (2012) called for replacing the notion of 'nature' of science (NOS) with 'features' of science (FOS). Matthews (2012) suggested shifting the focus from NOS to FOS in an effort to encompass a more inclusive range of ideas about science than would be possible than strictly following an epistemological emphasis, or focusing on scientific knowledge, as is the case with the "consensus view". The features proposed by Matthews include experimentation, idealization, models, values and socio-scientific issues, mathematization, technology, explanation, worldviews and religion, theory choice and rationality, feminism, realism and constructivism. Matthews (2012) believes that "we should have modest goals when teaching about FOS" (p. 20). Citing the intent of a major reform initiative like *Science for All Americans* (AAAS, 1989) and the *Perspectives of Science* (2007) course, Matthews argues that NOS teaching should focus on helping students get an appreciation of NOS ideas rather than gain "declarative knowledge of NOS." When this appreciation becomes a focus of instruction, the expanded list of features present viable ideas around which school science can be discussed from epistemic and historical points of view.

Matthews offers good reasons for why NOS ideas need to be broadened but does not present an explicit rationale for selecting these specific features of science and not others. The FOS features he has proposed resemble a disparate set of ideas some of which reflect epistemic aspects of science on the one hand (e.g. explanation,

## 1.2 Brief History of NOS in Science Education Research

theory choice and rationality), while others reflect a philosophical stance (e.g. feminism, realism and constructivism). In this sense, these features of science address different levels of organization of science and philosophy of science. In order to express these features in educational settings, it will be important to subsume these individual features of science under broader themes and then articulate the way in which they can be tapped to contribute to NOS discussions in a more coherent and wholesome way. But even then, it is debatable whether one of these themes, such as that pertaining to explicit reference of philosophical positions, is worthy of inclusion in the context of FOS-oriented school science.

An additional perspective is expressed by Allchin (2011), who has called for "reframing current NOS characterizations from selective lists of tenets to the multiple dimensions shaping reliability in scientific practice, from the experimental to the social, namely to Whole Science" (p. 518). Allchin argues that many items related to science as an enterprise, for instance, the role of funding, motivations, peer review, cognitive biases, fraud, and the validation of new methods, are absent in the so-called "consensus view" NOS list.

> The inventory may seem long and unwieldy, but (unlike the consensus list) it is unified by the theme of reliability. Items may also be easily organized: by following claims as they unfold in successively broadening contexts, from observational settings to public forums. Ironically, such a profile of reliability in scientific practice parallels potential sources of error, or error types, in science. We may need to inform students about all the ways scientific claims may fail, so that they understand how we prevent, mitigate, or accommodate potential error. Complete understanding of NOS, in this view, has both breadth and depth (completeness and proficiency). (Allchin, 2011, p. 524)

Allchin goes on to argue for reframing NOS to be sensitive to all dimensions of reliability in scientific practice:

> Whole Science, like whole food, does not exclude essential ingredients. It supports healthier understanding. Metaphorically, educators must discourage a diet of highly processed, refined "school science." Short lists of NOS features should be recognized as inherently incomplete and insufficient for functional scientific literacy. (Allchin, 2011, p. 524)

It may be helpful to place these arguments in historical perspective. Duschl and Grandy (2011) outline three phases in the development of twentieth century philosophy of science: an experiment-driven enterprise, a theory-driven enterprise and a model-driven enterprise. Each of these frameworks have led to a different conceptualization of the scientific method. For instance, the experiment-driven enterprise forms the foundation of logical positivism leading to the hypothetico-deductive conception of science. Such a "received view" of science is related to the traditional "scientific method" conceptualization with the following steps: (a) make observations, (b) formulate a hypothesis, (c) deduce consequences from the hypotheses, (e) accept or reject the hypothesis based on the observations. Duschl and Grandy propose that for a scientific method framework to be useful, it must incorporate the epistemic practices inherent in constructing and evaluating models because "theory-building and model-building practices provide the contexts where epistemic abilities, social skills and cognitive capacities are forged" (p. 8). They argue that the contemporary accounts of NOS in science education have not sufficiently addressed the dialectical processes that shape the role of theory, evidence, explanation and models that are involved in the development of scientific knowledge.

Furthermore, Duschl and Grandy (2011) stressed the significance of targeting understanding of the revision and modification of methods of inquiry as well as knowledge evaluation in science as learning outcomes in understanding the nature of science:

> One of the important findings from the science studies literature is that not only does scientific knowledge change over time, but so, too do the methods of inquiry and the criteria for the evaluation of knowledge change. (Duschl & Grandy, 2011, p. 17)

A key question in the characterization of NOS is "*Who decides for science education organizations and researchers the primarily philosophically based question of what are the tenets of the NOS?*" (Alters, 1997, p. 42). Alters, who carried out an empirical study on the perceptions and recommendations of philosophers of science, believes that philosophers of science should be brought into the picture not only to examine the different proposals about the NOS beliefs, but also to provide some guidance in establishing more precise criteria for the NOS. Osborne, Collins, Ratcliffe, Millar, and Duschl (2003) have surveyed a sample of 23 experts from diverse backgrounds that included science educators; scientists; historians, philosophers, and sociologists of science; experts engaged in work to improve the public understanding of science; and expert science teachers. The findings have revealed few themes on which the experts seemed to have some level of consensus. Five themes were subsumed under methods of science, two fell under the nature of scientific knowledge, and one fell under institutions and social practices of science. The authors acknowledge that "no one method and no one group of individuals can provide a universal solution as to what should be the essential elements of a contemporary science curriculum" (Osborne et al., 2003, p. 715).

The recent emergence of "Science Studies" in science education (e.g. Duschl, Erduran, Grandy, & Rudolph, 2006) extends such discussions on NOS to potentially include professionals in related foundational fields that can speak to the nature of science from a range of perspectives. "Science studies" is an interdisciplinary agenda drawing from history, philosophy, anthropology, and sociology of science as well as cognitive psychology and artificial intelligence to understand the nature of science. The relevance of "Science Studies" for science education has been argued on the basis that science classrooms are inherently inclusive of a range of features that call for characterization of science from a multitude of perspectives. The establishment in 2006 of a new section in the journal *Science Education* focusing on Science Studies and the subsequent growth of research in this area is an indicator of shifting expectations for nature of science to be more inclusive of interdisciplinary perspectives.

In their brief review of the recent developments in science studies, Duschl et al. (2006) illustrate the broadening of the perspectives on the nature of science from a narrow focus on the logic of scientific processes and conceptual outcomes of science to one that is more indicative of practices that scientists are engaged in:

Science studies now recognize that scientists do not just collect data, they design experiments to collect the data and they refine and interpret both the data and experiments. In each case what they could do and what they actually do are influenced by their motivations, by the social, informational, and technological resources available, and by the available alternative theories and models. Scientists also write, read, and argue about data, models, theories, and explanations, and in each case there are social and cultural contexts that influence their interpretations or their choice of statements. (Duschl et al., 2006, p. 963)

In an earlier discussion on the interdisciplinary accounts of science, Eflin et al. (1999) offer a brief taxonomy of the main issues in philosophy of science to help focus the debate for science education. They base this taxonomy on the following philosophical themes: Unity versus disunity of science, Demarcation, Realism versus Instrumentalism, Rationalism versus Historicism, Practice and Experiment versus Theory, and Feminist philosophy of science. Rather than appealing to philosophers as authorities, they suggest that science educators become acquainted with philosophical debates about science, and with the arguments and kinds of evidence used in favor of different positions that emerge from the above taxonomy of themes. It is in this spirit and recommendations that the key approach of this book is based on arguments and evidence from the philosophy of science. In the rest of the book, theoretical arguments based on Family Resemblance Approach (FRA) proposed by philosophers of science Irzik and Nola (2014, 2011) are reviewed, extended and applied to science education. The FRA approach will be detailed and discussed fully in Chap. 2.

It suffices to say that the choice of the FRA is based on the observation that it is broad and comprehensive enough to accommodate a variety of aspects of science including the epistemic, cognitive, and social aspects of science. Although we draw on Irzik and Nola's (2014) framework, we adapt it in significant ways. Irzik and Nola's framework is expanded and extended and hence, our interpretation of their approach may not in all instances correspond to how they envisioned FRA. For example, as part of the idea of "science as a social system", we introduce explicit attention to the political and financial dimensions of science, which are implied but not elaborated in their FRA. In re-conceptualizing the various epistemic, cognitive and social aspects of science, we have reconfigured the various notions to generate a framework that can be used in science education. In other words, our interpretation of FRA is not based exclusively on a philosophical analysis or synthesis but is informed by our knowledge of science education research and practice as well. Consequently, we offer in each chapter implications for teaching and learning that are neither contained nor implied by Irzik and Nola.

## 1.3 NOS in Curricular Context

A wide range of international policy documents (DfES/QCA, 2006; Education Commission, 2000; OECD, 2011) have highlighted the significance of the broader goal of scientific literacy for all students in secondary schooling. Curriculum reform efforts have concentrated on the teaching of science as a goal not only for the education of scientists but also for the general public. The key premise of these efforts is that in industrialized and democratic societies, as part of active citizenship, the public needs to be better equipped with scientific reasoning skills for informed decision-making about numerous issues ranging from climate change to genetic cloning. A particular aspect of the move for "scientific literacy for all" is the inclusion of themes such as NOS and the understanding of science in its socio-cultural context.

The shift from learning the conceptual outcomes of science towards inquiry-based approaches to science learning has been endorsed in various policy documents across the world, including the European Commission's "Rocard Report" on science education (Rocard, Csermely, Jorde, Lenzen, & Walberg-Henriksson, 2007) which recommends "a reversal of school science-teaching pedagogy from mainly deductive to inquiry-based methods provides the means to increase interest in science" (p. 12). A related trend has been the notion that science and society are no longer seen as separate actors where certain institutions have the monopoly for knowledge and other stakeholders (e.g. the public) as receivers of scientific knowledge. On the contrary, bridges with other stakeholders are being built and the relationship between scientific institutions and stakeholders is perceived to be interactive. In line with the Rocard Report, science communication is seen as *transaction* rather than *transmission*, the former involving "…a more symmetric, though not necessarily more equal, notion of communication. The starting point is that scientists and the public can learn from each other, that both have access to knowledge…". Hence "this transaction is an ongoing exchange of information, debate and knowledge that becomes an interaction" (pp. 50–51). However, the report states that "…the challenge remains to develop transaction modes of science communication. A further challenge is the shared construction of possible futures" (p. 11).

The contemporary arguments for the inclusion of NOS in science curriculum policy mirror earlier initiatives. For example, a crucial forerunner of science curriculum reform in the United States, Project 2061: *Science For All Americans*, a report prepared by the American Association for the Advancement of Science (1989) had articulated the view that understanding the nature of science, mathematics, and technology constitute one of four categories considered essential for all citizens in a scientifically literate society. Three principle components of the nature of science, as defined by AAAS are:

1. Scientific world view – the world is understandable, scientific ideas are subject to change, scientific knowledge is durable, and science cannot provide complete answers to all questions;
2. Scientific methods of inquiry – science demands evidence, science is a blend of logic and imagination, science explains and predicts, scientists try to identify and avoid bias, and science is not authoritarian; and
3. Nature of the scientific enterprise – science is a complex social activity, science is organized into content disciplines and is conducted in various institutions, there are generally accepted ethical principles in the conduct of science, and scientists participate in public affairs both as specialists and as citizens. (AAAS, 1989, pp. 25–31).

Curriculum documents from around the world have been advocating the incorporation of NOS in science education. For example in England and Wales, there have been renewed interest in the incorporation of themes that focus on knowledge construction as opposed to knowledge transmission. The "How Science Works" component of the Science National Curriculum (DfES/QCA, 2006) suggests the incorporation of aspects of NOS in various aspects of science teaching and learning. For instance, not only should students learn about coordination of evidence and explanation but also they should be communicating arguments.

The case of Hong Kong provides another example of recent international trends in the way that the science curriculum has been restructured. During 1970s–1990s, the economy of Hong Kong underwent a dramatic structural change from labor-intensive manufacturing to skill-intensive service industries that demands school leavers and university graduates to possess generic skills such as problem solving, investigative skills and self-learning ability. As Erduran and Wong (2013) discuss, such changes have resulted in the broadening the curriculum goals of science education from knowledge-focused goals to expanded ones covering the development of skills and attitudes (Education Commission, 2000). The curriculum in Hong Kong (Curriculum Development Council [CDC], 1998) encourages teachers to conduct scientific investigations in their classes, advocates scientific investigation as a desired means of learning scientific knowledge, and highlights the development of inquiry practices and generic skills such as collaboration and communication. As Erduran and Wong (2013) state, it is the first local science curriculum that embraced some features of nature of science (NOS), e.g. being "able to appreciate and understand the evolutionary nature of scientific knowledge" (CDC, 1998, p. 3), was stated as one of its broad curriculum aims. In the first topic, "What is science?", teachers are expected to discuss with students some features about science, e.g. its scope and limitations, some typical features about scientific investigations, including fair testing, control of variables, predictions, hypothesis, inferences, and conclusions. Such an emphasis on NOS was further reinforced in the revised secondary curricula (CDC, 2002). Scientific investigation continued to be an important component while the scope of NOS was slightly extended to include recognition of the usefulness and limitations of science as well as the interactions between science, technology, and society (STS).

The content of science curricula on NOS largely depends on epistemological orientation towards science. Various philosophical themes underscore the assumptions made by policy makers on what counts as science and to whom. In *Science Teaching: The Role of History and Philosophy of Science*, Michael Matthews (1994) explains how universalism presents science as a practice that cuts across all cultures, races, genders and religions. Although it is recognized that some aspects of culture influence science, Matthews considers that cultural influences do not determine the truth claims of science. Universalism displays science as an intellectual activity whose outcomes transcend human differences. Others such as Cobern and Loving (2001) explain how universalistic perspectives, such as the one called the "Standard Account" confer western science with a biased epistemological authority to decide what is truth. The way scientific knowledge is taught in schools is traditionally considered to help the individual not only in the acquisition of a desired economic status, but also provides power among the men and women who own it

(Cobern & Loving, 2001). Cobern and Loving (2001) explain this cultural hegemony of science in terms of 'scientism', the view that there are no real limits to the competence of science, that science is the vehicle through which we humans can achieve or know about reality. Other authors have reflected on 'scientism' indicating that it leads us to the belief that "[t]here is nothing outside the domain of science, nor is there any area of human life to which science cannot successfully be applied" (Stenmark, 1997, p. 15).

Another theme that has influenced the development of curricular frameworks on NOS is constructivism, which positions students as active protagonists in the building and acquisition of their science knowledge. There is substantial research conducted on constructivism by philosophers of science as well as science educators, with the overall aim of articulating how scientific knowledge is developed. In other words, constructivism debate is relevant for the NOS discussions because it speaks to how scientific knowledge is constructed. Constructivism can also be relevant for NOS due to the ontological implications of its various versions. The educational process from a constructivist orientation positions teachers as guides who facilitate students' re-construction of knowledge, while students build upon their own ideas through active participation to develop a deeper conceptual understanding of science. Despite a general consensus among science curriculum developers that science curriculum should take a constructivist approach to learning, disagreements exist. Part of the reason for disagreement is that constructivism has many variations. Geelan (1997) classified six forms of constructivism: (a) Personal constructivism, represented by scholars such as Piaget, (b) Radical constructivism whose major representative is von Glasersfeld, (c) Social constructivism, as described by Solomon, (d) Social constructionism, explained by Gergen's philosophy, (e) Critical constructivism represented by Taylor's work, and (f) Contextual constructivism which is explained by contemporary academics such as Cobern.

Michael Matthews classified constructivism in terms of three categories: (a) educational, personal and social constructivism, (b) philosophical constructivism, originating in Kuhn's work, and (c) sociological constructivism, which he explains in terms of the Sociology of Scientific Knowledge, making reference to Edinburgh "Strong Programme" (Matthews, 1997). Gürol Irzik (2001) on the other hand, distinguished between different forms of constructivism. He finds constructivism of the educational and social kind to be cognitive, epistemic, semantical and metaphysical. Irzik points at the flaws of these different views, stressing the ambiguity that the use of this vocabulary produces inevitably leading to the loss of the right or wrong idea in science education. In addition, he recognizes that constructivism is a reaction to the didactical, teacher-oriented "transmission model" of education (traditional education), and therefore he acknowledges that some constructivist ideals have been beneficial to science education.

Overall the various curricular goals across the world and the plethora of philosophical underpinnings, particularly in reference to variations of constructivism, have had a direct link to how NOS is defined for science teaching and learning at the level of the classroom. Despite the tensions and the diversity of opinions surrounding curriculum reform, it is worthwhile to remember that the scientifically literate person will be more likely to accept change in scientific ideas if he or she possess a better understanding of the nature of scientific inquiry (Duschl, 1990). Without an adequate understanding of the warrants or reasons scientists use to change methods, beliefs and processes, it is probable that people will not acknowledge scientists' views as rational and as products of a process.

## 1.4 Key Contributions of the Book

The main aims of this book are to broaden the discussions on NOS in science education, and to open up new spaces for research and development in science curriculum, teaching, learning and teacher education. In broadening the definition of 'science' to be inclusive of the epistemic, cognitive and social aspects of science in a coordinated fashion, a wider range of learners can potentially be engaged in science. In other words, conceptualizations of science that are inclusive of aspects of science that are not typically covered in school science can potentially be motivating and interesting to a more diverse student population. The approach taken in the book is theoretical in nature drawing on arguments primarily from evidence from philosophy of science literature. The intention is to provide a sound rationale for what 'science' is about and consider the implications of how it could be taught. In drawing on aspects of science that are typically underrepresented in the characterizations of NOS (e.g. political and financial aspects), we hope to provide an improved representation of science in school science. Although other researchers have also provided similar arguments for being inclusive of various aspects of science in science education, the collective and holistic treatment of science presented here is unique. The coverage of the book ranges from the aims and values to methods, practices and contexts of science in a coordinated fashion. Furthermore the book articulates not only the various facets of science but also clarifies the nuances in the way that different branches of science might deal with particular research problems. The latter emphasis on the domain-specific nature of NOS in different branches of science is an understudied area of research. Admittedly, due to the theoretical nature of the work, the level of detail for practical applications may at times be insufficiently covered. A comprehensive and detailed account of the pedagogical, curricular and teacher education aspects of the work are considered as a next step and for this purpose, some recommendations are provided in Chap. 8.

As indicated earlier, the main orientation to the book is based on Irzik and Nola's (2014) Family Resemblance Approach (FRA), which we describe in Chap. 2. We elaborate, critique, extend and apply this approach to science education. In so doing, we are using our value judgments and pedagogical filters to select philosophical content that is most relevant to educational contexts. As science educators with experience as teachers, teacher educators as well as researchers, we care deeply about science learning. Hence our project is not to teach NOS for NOS sake but to teach it in a way that makes science learning most wholesome and thorough, aiming for true scientific proficiency.

As the following chapters illustrate, the key argument relies on the detailing of different categories that contribute to the characterization of science. A set of visual tools are proposed to enhance understanding of some complex dimensions of science. It is widely reported that visualization can facilitate teaching and learning of science (e.g. Gilbert, 2005; Gilbert, Reiner & Nakhleh, 2008; Johnson-Laird, 1998; Wu, Krajcik, & Soloway, 2001). For example, Phillips and colleagues argue that "visualizing objects assist in explaining, developing, and learning concepts in the field of science" (Phillips, Norris, & Macnab, 2010, p. 63). Incorporation of visual tools is novel to research on NOS in science education. Extant perspectives on the consensus view of NOS (e.g. Abd-El-Khalick & Lederman, 2000; Schwartz, Lederman, & Crawford, 2004) or alternative proposals (e.g. Allchin, 2011; Duschl & Grandy, 2013; Irzik & Nola, 2014) do not provide visual representations of the various epistemic, cognitive and social dimensions of science in a way that can be transformed for pedagogical use. Visual representations around the aims and values, contexts as well as practices of science can help researchers, teachers and learners in making sense of science in an organized fashion. The images may need to be extended, adapted and translated for various uses, depending on the readers' purposes. This generative aspect of the images is a strength of the book in providing science educators tools that can guide inquiry, reflection and learning.

A consistent theme across the book is the application of the respective ideas to a curriculum policy context. For the most part, issues and examples are related to the *Next Generation Science Standards* (*NGSS*) (NGSS Lead States 2013) primarily due to the novelty of these standards at the time of the writing of the book as well as their anticipated impact in the United States and around the world. Relating the revised account of NOS to the *NGSS* and the *Framework for K-12 Science Education,* which set out the vision for developing these standards (NRC, 2012), can provide useful guidelines for applying the characterization of science described in this book to curriculum policy documents in other countries. It should be noted that choosing to illustrate the relevance of our ideas in the context of *NGSS* does not imply that we are necessarily endorsing NGSS. Rather, as we are mindful of the necessity to link science education research to a contemporary and relevant policy document, an example reference is included.

Although we make the case for a holistic characterization of science for science education, we are cautious about the extent to which our discussion or any discussion among the science education research community can represent the work of philosophers of science or indeed science itself. In trying to build up an image of science through a wide range of orientations that rely on the epistemic, cognitive and social dimensions of science, it is possible that particular ideas or characterizations of science are omitted. It is conceivable that attention can be given to even wider range of disciplinary characterizations of science from, for instance, anthropological, sociological and cultural accounts of science. In this sense, our depiction is not unlike a set of projections that try to approximate a complex object or phenomenon and thus relies on the accumulation and interpretation of a set of projections. This issue can be likened to an image produced by Trevor Shannon, a young mechanical engineer based in California (see Fig. 1.1).

## 1.4 Key Contributions of the Book

**Fig. 1.1** Shadows of an object by Trevor Shannon (Obtained with permission through personal communication, July 28, 2013)

Consider, metaphorically, science as an object of interest being represented by the object in the picture. The different impressions about what science is are pieced together from the shadows cast by a light source assuming it is not possible for the viewer to see the object directly. Notice that there could be an infinite number of shadows created from one or more light sources. Depending on the perspective by which a 'shadow' is cast on science (i.e. epistemic, cognitive, social, institutional) different representations may emerge that are not entirely based on the "full picture" of a particular branch of science such as biology, chemistry and physics, let alone science as a unitary endeavor. Even with best of efforts, the union and compilation of shadows from multiple perspectives will not directly help reconstitute the nature of science. A process of inferences and reconfiguration will need to be operating in order to best estimate what science involves. Reconceptualizing NOS for science education essentially adds a few more shadows, so to speak, through which the nature of science can be better approximated but it does not necessarily complete the picture. The expanded Family Resemblance Approach enables the approximation of the various domain-general (e.g. the value of objectivity of scientific claims) and domain-specific aspects of science (e.g. role of experimentation in astronomy versus physics) such that we can begin to discuss how the various components of science (e.g. values, methods) can be characterized in a fine-grained way. Thus, the extension, elaboration and addition of the often-neglected components of science in a more holistic fashion is a step forward in contextualizing science and making its nature more visible and comprehensible to learners.

Having outlined the main purpose, it is worthwhile to clarify what this book is not about. The book is not exhaustive of the topics, issues and debates from philosophy of science nor other foundational disciplines. Furthermore, the purpose is not to provide an instructional manual for applying the theoretical ideas derived from philosophers' work. We do not intend to portray an impression that debates around the philosophical issues are settled. Rather, we invite a community of stakeholders in science education to engage with these ideas. The community includes but is not limited to researchers, curriculum policy makers, teachers and teacher educators. In sum, this book reconceptualizes the nature of science by developing a theoretical framework that is flexible and inclusive, and illustrates the pertinence of this framework for science education.

# References

AAAS. (1989). *Science for all Americans*. Washington, DC: American Association for the Advancement of Science.
Abd-El-Khalick, F. (2012). Examining the sources for our understandings about science: Enduring conflations and critical issues in research on nature of science in science education. *International Journal of Science Education, 34*(3), 353–374.
Abd-El-Khalick, F., Bell, R. L., & Lederman, N. G. (1998). The nature of science and instructional practice: Making the unnatural natural. *Science Education, 82*(4), 417–436.
Abd-El-Khalick, F., & Lederman, N. G. (2000). Improving science teachers' conceptions of nature of science: A critical review of the literature. *International Journal of Science Education, 22*(7), 665–701.
Ackerson, V., & Donnelly, L. A. (2008). Relationships among learner characteristics and preservice teachers' views of the nature of science. *Journal of Elementary Science Education, 20*(1), 45–58.
Allchin, D. (2011). Evaluating knowledge of the nature of (whole) science. *Science Education, 95*(3), 518–542.
Allchin, D. (2013). *Teaching the nature of science: Perspectives and resources*. St. Paul, MN: SHiPs.
Alters, B. J. (1997). Whose nature of science? *Journal of Research in Science Teaching, 34*(1), 39–55.
Booth, T., & Ainscow, M. (2000). *Index on inclusion*. Bristol, UK: Centre for Studies on Inclusive Education.
CDC. (2002). *Physics/Chemistry/Biology curriculum guide* (secondary 4–5). Hong Kong: Curriculum Development Council. Retrieved August 15, 2011, from http://www.edb.gov.hk/index.aspx?nodeID=2824&langno=1
Cobern, W. W., & Loving, C. C. (2001). Defining "science" in a multicultural world: Implications for science education. *Science Education, 85*, 50–67.
Curriculum Development Council (CDC). (1998). *Science syllabus for secondary 1–3*. Hong Kong: CDC.
Chang, Y., Chang, C., & Tseng, Y. (2010). Trends of science education research: An automatic content analysis. *Journal of Science Education and Technology, 19*, 315–332.
Clough, M. P. (2007, January). Teaching the nature of science to secondary and post-secondary students: Questions rather than tenets. *The Pantaneto Forum*, Issue 25, http://www.pantaneto.co.uk/issue25/front25.htm
Conant, J. (1961). *Science and common sense*. New Haven, CT: Yale University Press.
Cooley, W., & Klopfer, L. (1963). The evaluation of specific educational innovations. *Journal of Research in Science Teaching, 1*, 73–80.
Cotham, J., & Smith, E. (1981). Development and validation of the conceptions of scientific theories test. *Journal of Research in Science Teaching, 18*(5), 387–396.
Department for Education and Skills and Qualifications and Curriculum Authority. (2006). *Science. The national curriculum for England*. London: HMSO.

# References

Driver, R., Leach, J., Millar, R., & Scott, P. (1996). *Young people's images of science*. Buckingham, UK: Open University Press.

Duschl, R. (1990). *Restructuring science education: The importance of theories and their development*. New York: Teachers College Press.

Duschl, R., Erduran, S., Grandy, R., & Rudolph, J. (2006). Guest editorial: Science studies and science education. *Science Education, 90*(6), 961–964.

Duschl, R., & Grandy, R. (2011). Demarcation in science education: Toward an enhanced view of scientific method. In R. Taylor & M. Ferrari (Eds.), *Epistemology and science education: Understanding the evolution vs. intelligent design controversy* (pp. 3–19). New York: Routledge.

Duschl, R., & Grandy, R. (2013). Two views about explicitly teaching nature of science. *Science & Education, 22*, 2109–2139.

Education Commission. (2000). *Learning for life, learning through life: Reform proposals for the 780 education system in Hong Kong*. Hong Kong: Education Commission.

Eflin, J. T., Glennan, S., & Reisch, G. (1999). The nature of science: A perspective from the philosophy of science. *Journal of Research in Science Teaching, 36*(1), 107–117.

Erduran, S., & Wong, A. (2013). Science curriculum reform on "Science for All" across national contexts: Case studies of curricula from England and Hong Kong. In N. Mansour & R. Wegeriff (Eds.), *Science education for diversity in knowledge society* (pp. 179–201). Dordrecht, The Netherlands: Springer.

Ford, M. (2008). 'Grasp of practice' as a reasoning resource for inquiry and nature of science understanding. *Science & Education, 17*, 147–177.

Geelan, D. R. (1997). Epistemological anarchy and the many forms of constructivism. *Science & Education, 6*(1–2), 15–28.

Gilbert, J. (Ed.). (2005). *Visualisation in science education*. Dordrecht, The Netherlands: Springer.

Gilbert, J. K., Reiner, M., & Nakhleh, M. (Eds.). (2008). *Visualization: Theory and practice in science education*. Dordrecht, The Netherlands: Springer.

Grandy, R., & Duschl, R. (2007). Reconsidering the character and role of inquiry in school science: Analysis of a conference. *Science & Education, 16*(1), 141–166.

Irzik, G. (2001). Back to basics: A philosophical critique of constructivism. *Science & Education, 20*, 157–175.

Irzik, G., & Nola, R. (2011). A family resemblance approach to the nature of science. *Science & Education, 20*, 591–607.

Irzik, G., & Nola, R. (2014). New directions for nature of science research. In M. Matthews (Ed.), *International handbook of research in history, philosophy and science teaching* (pp. 999–1021). Dordrecht, The Netherlands: Springer.

Johnson-Laird, P. N. (1998). Imagery, visualization, and thinking. In J. E. Hochberg (Ed.), *Perception and cognition at century's end* (pp. 441–467). San Diego, CA: Academic.

Kimball, M. (1968). Understanding the nature of science: A comparison of scientists and science teachers. *Journal of Research in Science Teaching, 5*, 110–120.

Klopfer, L. (1969). The teaching of science and the history of science. *Journal of Research in Science Teaching, 6*, 87–95.

Lederman, N. G. (1992). Students' and teachers' conceptions of the nature of science: A review of the research. *Journal of Research in Science Teaching, 29*(4), 331–359.

Lederman, N. G. (2007). Nature of science: Past, present, future. In S. Abell & N. Lederman (Eds.), *Handbook of research on science education* (pp. 831–879). Mahwah, NJ: Lawrence Erlbaum.

Lederman, N. G., Abd-El-Khalick, F., Bell, R. L., & Schwartz, R. S. (2002). Views of nature of science questionnaire (VNOS): Toward valid and meaningful assessment of learners conceptions of nature of science. *Journal of Research in Science Teaching, 39*(6), 497–521.

Matthews, M. (1994). *Science teaching: The role of history and philosophy of science*. New York: Routledge.

Matthews, M. (1997). Introductory comments on philosophy and constructivism in science education. *Science & Education, 6*, 5–14.

Matthews, M. (2012). Changing the focus: From nature of science (NOS) to features of science (FOS). In M. S. Khine (Ed.), *Advances in nature of science research* (pp. 3–26). Dordrecht, The Netherlands: Springer.

Matthews, M. (Ed.). (2014). *Handbook of research on history, philosophy and sociology of science*. Dordrecht, The Netherlands: Springer.

McComas, W. F. (Ed.) (1998). The nature of science in science education: Rationales and strategies. Dordrecht, Netherlands: Kluwer (Springer) Academic Publishers.

McComas, W. F., Clough, M. P., & Almazroa, H. (1998). The role and character of the nature of science in science education. *Science & Education, 7*(6), 511–532.

McComas, W. F., & Olson, J. K. (1998). The nature of science in international science education standards documents. In W. F. McComas (Ed.), *The nature of science in science education: Rationales and strategies* (pp. 41–52). Dordrecht, The Netherlands: Kluwer Academic.

National Research Council. (2012). A framework for K-12 science education: Practices, crosscutting concepts, and core ideas. Washington, DC: National Academies Press.

NGSS Lead States. (2013). Next generation science standards: For states, by states. Appendix H. Retrieved from http://www.nextgenscience.org/next-generation-science-standards

OECD. (2011). *Education at a glance 2011: OECD indicators*. OECD Publishing. doi:10.1787/eag-2011-en

Osborne, J., Collins, S., Ratcliffe, M., Millar, R., & Duschl, R. (2003). What "ideas-about-science" should be taught in school science? A Delphi study of the expert community. *Journal of Research in Science Teaching, 40*(7), 692–720.

Phillips, L. M., Norris, S. P., & Macnab, J. S. (2010). *Visualization in mathematics, reading and science education*. Dordrecht, The Netherlands: Springer.

Rocard, M., Csermely, P., Jorde, D., Lenzen, D., & Walberg-Henriksson, H. (2007). *Rocard report of the European Commission. Science education now. A renewed pedagogy for the future of Europe*. Brussels, Belgium: European Commission. ISBN 978-92-79-05659-8.

Rubba, P., & Anderson, H. (1978). Development of an instrument to assess secondary students' understanding of the nature of scientific knowledge. *Science Education, 62*(4), 449–458.

Ryder, J., Leach, J., & Driver, R. (1999). Undergraduate science students' images of science. *Journal of Research in Science Teaching, 36*(2), 201–220.

Sandoval, W. A. (2005). Understanding students' practical epistemologies and their influence on learning through inquiry. *Science Education, 89*(4), 634–656.

Schwartz, R. S., Lederman, N. G., & Crawford, B. A. (2004). Developing views of nature of science in an authentic context: An explicit approach to bridging the gap between nature of science and scientific inquiry. *Science Education, 88*(4), 610–645.

Showalter, V. (1974). What is unified science education? Program objectives and scientific literacy (Part 5). *Prism II, 2*(3), 1–3.

Smith, M. U., Lederman, N. G., Bell, R. L., McComas, W. F., & Clough, M. P. (1997). How great is the disagreement about the nature of science: A response to Alters. *Journal of Research in Science Teaching, 34*(10), 1101–1103.

Stenmark, M. (1997). What is scientism? *Religious Studies, 33*(1), 15–32.

Wu, H. K., Krajcik, J. S., & Soloway, E. (2001). Promoting understanding of chemical representations: Students' use of a visualization tool in the classroom. *Journal of Research in Science Teaching, 38*(7), 821–842.

Yacoubian, H. (2012). *Towards a philosophically and a pedagogically reasonable nature of science curriculum*. Doctoral dissertation. Retrieved from http://era.library.ualberta.ca/public/view/item/uuid:9b2d52c1-607a-420b-8447-54c82ae14a72

# Chapter 2
# Family Resemblance Approach to Characterizing Science

The chapter draws on the Family Resemblance Approach (FRA) to inform characterisations of nature of science in science education. The components of the FRA are described and a rationale is provided for its relevance in science education. The FRA can provide a fresh new perspective on how science can be conceptualized in general and how such conceptualisation can be useful for teaching and learning of science in particular. The FRA is described and extended being mindful to have sufficient context and content for it to be of use for science education purposes. Izrik and Nola's (2014) depiction of FRA, which describes components of science in terms of categories subsumed under epistemic, cognitive and social systems is used. However, these authors framework does not provide an extensive discussion. Indeed, the description of their categories is rather brief. The aim of the chapter is to build on the FRA itself and explore its potential for use in science education. In applying the FRA to science education, Irzik and Nola's philosophical model is developed into a functional framework for instructional and learning purposes throughout the rest of this book. In particular, the authors' linguistic and textual account is transformed into a visual representation that highlights the need for a dynamic and interactive tool representing science in a holistic account. The transformed FRA informs the content and structure of the chapters.

## 2.1 Introduction

As discussed in Chap. 1, there are multiple ways in which nature of science has been defined, and various arguments advanced to support different formulations. We take the position that nature of science in its broader sense encapsulates a range of practices, methodologies, aims and values, and social norms that have to be acknowledged when teaching science. Restricting nature of science in the context of school science to a limited set of ideas about the nature of scientific knowledge unduly results in limited attention to other core factors that influence the formation and validation of scientific claims. For example, not understanding the way in which cultures of science are constituted and how these cultures contribute to the development of scientific knowledge will result in a rather narrow understanding of science as a human endeavor.

Irzik and Nola (2011a, 2014) attempt to address the unity of science without sacrificing its diversity by pursuing a Family Resemblance Approach. Basing their notion of family resemblance on Wittgenstein's work, they present their scheme as an alternative to the consensus view, arguing that it is "more comprehensive and systematic" (Irzik & Nola, 2014, p. 1000). The advantage of using the FRA to characterize a field of science is that it allows a set of broad categories to address a diverse set of features that are common to all the sciences. This is particularly useful in science, whereby all subdisciplines share common characteristics but none of these characteristics can define science or demarcate it from other disciplines. For instance, Irzik and Nola (2014) present the example of observation (i.e. human or artificial through the use of detecting devices) and argue that even though observing is common to all the sciences, the very act of observing is not exclusive to science and therefore does not necessarily grant family membership. The same applies to other practices such as inferring and data collecting, which are shared by science fields but their use is not necessarily limited to them.

The family resemblance model of nature of science conceptualizes science in terms of a cognitive-epistemic and a social-institutional system. The analytical distinctions Irzik and Nola make are meant to "achieve conceptual clarity, [and] not [serve] as a categorical separation that divides one [dimension] from the other. In practice, the two constantly interact with each other in myriad ways" (Irzik & Nola, 2014, p. 1003). This is a critical distinction to uphold in this chapter as well as the rest of the book. Science as a cognitive-epistemic system encompasses processes of inquiry, aims and values, methods and methodological rules, and scientific knowledge, while science as a social-institutional system encompasses professional activities, scientific ethos, social certification and dissemination of scientific knowledge, and social values.

Within the cognitive-epistemic system, Irzik and Nola discuss four categories[1] described briefly as follows. The processes of inquiry considered in this scheme refer to types of activities that are rather familiar to science educators. They include activities like "posing questions (problems), making observations, collecting and classifying data,

---

[1] In the rest of the book we will use the term 'category' to denote the key components of science as a cognitive-epistemic and social-institutional system (see Table 2.1). In emphasizing the pedagogical applications and implications of the FRA framework, we will refer to 'epistemic', 'cognitive' and "social-institutional" aspects. At times, for the sake of brevity, we will collapse "social-institutional" aspects into 'social' or "social context".

designing experiments, formulating hypotheses, constructing theories and models, comparing alternative theories and models" (Irzik & Nola, 2014, p. 1007).

Aims and values refer to a set of aims in the sense that the products of scientific activity are desired to fulfill them. Aims and values include some obvious ones "such as prediction, explanation, consistency, simplicity and fruitfulness" (Irzik & Nola, 2014, p. 1007). Aims also include viability, testability, and empirical adequacy that function both as aims and values, and at times they function as shared criteria that play a significant role in theory choice.

Methods and methodological rules refer to the variety of systematic approaches and the rules that scientists use to ensure that they yield reliable knowledge. Included in these methods are different strategies such as inductive, deductive and abductive reasoning. Equally important to the methods are the set of methodological rules that guide their use. Examples of methodological rules are such statements as: "other things being equal choose the theory that is more explanatory," "use controlled experiments in testing casual hypotheses," and "in conducting experiments on human subjects always use blinded procedures" (Irzik & Nola, 2014, p. 1009). Scientific knowledge refers to the 'end-products' of scientific activity that culminate in "laws, theories, models as well as collection of observational reports and experimental data" (Irzik & Nola, 2014, p. 1010). Reference to end products is focused on the epistemic and cognitive aspects of these entities, how they become established, and what differentiates them from one another.

Within the conception of science as a social-institutional system, Irzik and Nola (2014) offer four categories that include professional activities, social and ethical norms, community aspects of science work, and the relationships of science with technology and society. Irzik and Nola are quick to admit that these categories are not exhaustive and that this may not be necessarily the best way to describe the social aspects of science. The shift in their original conception from sole focus on cognitive aspects of science (Irzik & Nola, 2011a) to adding one category of social context (Irzik & Nola, 2011b) to including four categories embedded under science as a social-institutional system creates more balance between the cognitive-epistemic and the social-institutional factors. This balance reflects the complex nature of science. It is also relevant to the broader goals of science education, as will be demonstrated throughout the book.

A brief description of the four categories under the social-institutional dimension follows. Professional activities refer to activities that scientists perform in order to communicate their research, such as attending professional meetings to present their findings, writing manuscripts for publications and developing grant proposals to obtain funding. Scientific ethos refers to the set of norms scientists follow in their own work and their interactions with one another. These include Mertonian norms (i.e. universalism, organized skepticism, disinterestedness, and communalism) as well as other ethical norms elaborated by Resnik (2007). The latter include things such as honesty and respect for research subjects and the environment. The social certification and dissemination of scientific knowledge refers to the peer review process, which tends to work as a *"social quality control* over and above the *epistemic control* mechanisms that include testing, evidential relations, and methodological consideration" (Irzik & Nola, 2014, p. 1014). The social values of science refer values such as "freedom, respect for the environment, and social utility broadly understood to refer to improving people's health and quality of life as well as to contributing to economic development" (Irzik & Nola, 2014, p. 1014) (Table 2.1).

These categories are not mutually exclusive entities but are complementary in the sense that they target different dimensions of the scientific enterprise. They are identified in separate categories to allow a more detailed analysis. Given the complexity of the scientific enterprise, it is helpful to disentangle some of its components, especially those that constitute commonalities across different domains. Irzik and Nola (2011a, 2011b; 2014) note that even though the processes, aims and values, methods and methodological rules, knowledge claims and the four aspects of the social institutional system may differ across science domains, there is enough resemblance along these categories within and across domains that make them recognizable as scientific.

Irzik and Nola (2014) describe the Family Resemblance Approach itself as follows:

> Consider a set of four characteristics {A, B, C, D}. Then one could imagine four 440 individual items which share any three of these characteristics taken together such as (A&B&C) or (B&C&D) or (A&B&D) or (A&C&D); that is, the various family resemblances are represented as four disjuncts of conjunctions of any three properties chosen from the original set of characteristics. This example of a polythetic model of family resemblances can be generalised as follows. Take any set S of n characteristics; then any individual is a member of the family if and only if it has all of the n characteristics of S, or any (n-1) conjunction of characteristics of S, or any (n-2) conjunction of characteristics of S, or any (n-3) conjunction of characteristics of S and so on. How large n may be and how small (n-x) may be is something that can be left open as befits the idea of a family resemblance which does not wish to impose arbitrary limits and leaves this to a 'case by case' investigation. In what follows we will employ this polythetic version of family resemblance (in a slightly modified form) in developing our conception of science. (Irzik & Nola, 2014, p. 1011)

They then proceed to argue that there are characteristics common to all sciences and some that are rather specific in emphases to particular sciences. For example, sciences share such practices as collecting data and making inferences. Other features of activities of science such as experimentation, however, might be differentiated. Irzik and Nola (2014) give the example of astronomy and earth sciences. These domains cannot possibly rely on experiments as celestial bodies cannot be manipulated. Likewise, earthquakes cannot be manipulated in the experimental sense. The authors situate the Family Resemblance Approach further by providing a disciplinary approach:

> Let us represent data collection, inference making, experimentation, prediction, hypothetico-deductive testing and blinded randomised trials as D, I, E, P, H and T, respectively. Then we can summarise the situation for the disciplines we have considered as follows:
>
> Astronomy = {D,I,P,H};
> Particle physics = {D,I,E,P,H};
> Earthquakescience = {D,I,P',H};
> Medicine = {D,I,P'',E,T}, where P'and P'' indicate differences in predictive power as indicated.

## 2.1 Introduction

**Table 2.1** Family resemblance approach (Irzik & Nola, 2014, p. 1009)

| Science | | | | | | | |
|---|---|---|---|---|---|---|---|
| Science as cognitive-epistemic system | | | | Science as a social-institutional[a] system | | | |
| 1 | 2 | 3 | 4 | 5 | 6 | 7 | 8 |
| Processes of inquiry | Aims and values | Methods and methodological rules | Scientific knowledge | Professional activities | Scientific ethos | Social certification and dissemination of scientific knowledge | Social values |

[a]The Table in Irzik and Nola (2014, p. 1009) does not reference to 'Institutional'. However the corresponding aspect discussed in their paper is "Social-Institutional System" as a section heading. Therefore we include the word 'institutional' in our reproduction of the Table

Thus, none of the four disciplines has all the six characteristics, though they share some of them. With respect to other characteristics, they partially overlap, like the members of closely related extended family. In short, taken altogether, they form a family resemblance.

Overall, The FRA provides an account where the domain-general and domain-specific aspects of science can be articulated. Illustrating the interplay between family resemblance features and how they get expressed in domain-specific contexts across science disciplines are addressed throughout the book.

## 2.2 Justifying the Family Resemblance Approach

One of the appealing aspects of the FRA is its ability to consolidate the epistemic, cognitive and social aspects of science in a wholesome, flexible, descriptive but non-prescriptive way. FRA provides focus zones that support the discussion of critical elements about science which can potentially be fruitful for science educators as well as teachers and students. It creates a much-needed space for conversation and dialog about science in a comprehensive way. It is this invitation to dialog that has intrigued us and provided us a foundational place to develop and expand what Irzik and Nola (2011a, 2011b, 2014) originally argued. As philosophers, they have presented a compelling justification for their framework. Their account is broad enough to accommodate further development and expansion. As science educators, we recognize in their framework a comprehensive organizational scheme that enables us to unpack the complex ideas that we judge worthy of expansion and application in science education.

Another advantage to the FRA is that it is an expansive framework that incorporates many components of existing nature of science frameworks. To elaborate this idea, two existing frameworks are considered, the consensus view and the features of science view, the latter intended to be a revisionary account of nature of science in science education. The components of three frameworks are aligned in Table 2.2 to illustrate how ideas from the consensus view and the FOS view relate to the FRA. The notation with the question mark (?) refers to instances where a comparable concept is either not explicitly present or could not be identified. Only a small set of ideas that represent philosophical positions such as constructivism, realism and feminism under the FOS approach are not directly addressed in the FRA because, as explained earlier, the FRA takes a neutral stance towards these positions. One could argue that these philosophical stances are constituted within the articulation of the eight categories that Irzik and Nola (2014) discuss. However, their work on FRA does not explicitly address these positions. The FRA framework appears to subsume all the individual components of the consensus and FOS frameworks.

Of note in this comparison is the difference in orientation afforded by the FRA in comparison to the consensus approach to teaching NOS. The FRA addresses a higher

## 2.2 Justifying the Family Resemblance Approach

level of organization involving a *class of ideas* approximating common characteristics. In contrast, the consensus view addresses *individual ideas* about science. For example, the FRA refers to scientific knowledge as a key cognitive epistemic category about science. In contrast the NOS consensus view distinguishes between scientific theories and laws. The former (i.e. scientific knowledge) is a class of ideas whereas the latter is an individual idea within that class. This is a fundamental difference between these two approaches. In our view, the higher level of organization in the FRA is precisely its strength as it lends itself to flexible exploration of those aspects about science that are most relevant to target science content. Ultimately, the purpose of the FRA as applied in educational settings is neither to teach students individual ideas nor to teach them specific philosophical doctrines about science but rather to promote holistic and contextualized understanding of science.

As Table 2.2 illustrates, FRA seems to capture a meta-level characterization of the key categories related to science in a broad sense. In other words, the FRA is more inclusive of various aspects in its depiction of science. It is the holistic, inclusive, diverse and comprehensive and meta-level conceptualization of FRA that has been appealing to us as science educators. Having awareness of a wider range of NOS issues does not necessarily mean that the curriculum, the teachers and the students will now be burdened with having to cover them all at once. The framework mainly invites selecting those issues about science that are of immediate relevance to the big ideas that are already under study. It alerts us to the missing components about the nature of science in science education such that we could make intelligent decisions about which aspect to prioritize when and for what purpose. Furthermore, having a more diverse representation of science has potentially more appeal to a wider range of students. For example, students who may not necessarily be drawn to the epistemic dimensions of science, may now find more motivation and interest in the social-institutional aspects of science. Hence, FRA approach potentially can be more inviting to learners. Arguably, some of the categories represented in the FRA may not conventionally be familiar to science teachers. We envisage this conversation to be the beginning of a new territory of professional development as well as research in science education. As illustrated in subsequent chapters, particularly in Chap. 8, there are also potentially fruitful spaces for policy makers in considering the often-neglected aspects of nature of science in the science curriculum.

Apart from a comprehensive set of categories about the cognitive-epistemic and social-institutional aspects of science, "family resemblance" enables the articulation of science through a set of comparisons between the different branches of science, thus allowing the consideration of domain-general as well as domain-specific set of characteristics of science. The "family resemblance" theme provides a much needed coherence to how we can envisage science from a more holistic perspective. In other words, while individual components from the particular eight categories might have been captured in other depictions of nature of science, these individual components can remain rather disconnected without an overarching and cohesive theoretical framework. The consequence of such lack of coherence between the different categories of science can potentially lead to restricted understanding about science.

**Table 2.2** Comparative overview of Nature of Science (NOS) consensus view, Features of Science (FOS) approach and the Family Resemblance Approach (FRA)

| NOS consensus view | Features of science approach | Family resemblance approach |
|---|---|---|
| Rationality Objectivity/Subjectivity | Lists: Theory choice and rationality which involve a set of aims and values | Includes scientific aims and values that subsume rationality and theory choice as an aim and value |
| ? | Lists practices that include: Experimentation, Idealization, Technology, Explanation, Mathematization | Includes nature of scientific practices pertaining to observation, experimentation, classification and so on |
| Focuses on the idea that scientists use many methods: no one scientific method | ? | Methodologies and methodological rules |
| Distinguishes between scientific theories and laws; observations and inferences; Focuses on tentativeness | Includes Models | Scientific knowledge: Epistemic-cognitive aspects of models, theories, laws and explanations and aspects pertaining to them such as knowledge revision |
| Highlights cultural embeddedness | Includes Values and socio-scientific issues; Worldviews and religion-Values and socio-scientific issues | The expanded social context recognizes cultural embeddedness and societal and religious values |
| Includes Creativity | ? | Creativity is a psychological component that characterizes aims and methods, practices, methods, and scientific knowledge. It in implicit in the FRA |
| ? | Includes the following philosophical positions: Realism, Constructivism, Feminism | The FRA does not make a commitment to any of these positions. In this sense, it is philosophically neutral |

Often in school science, it is indeed observed that students are introduced to rather discrete set of features of the nature of science without a meta-level understanding of how these discrete features relate to one other. The "family resemblance" approach has the potential to inform and generate more pedagogically, cognitively, and epistemically sound models of nature of science for science education.

## 2.3 Extending the Family Resemblance Approach

As mentioned earlier, one of the advantages of the FRA is that it lends itself to further development and to incorporation of related ideas. In order to keep the terminology clear, there are specific instances where we have intentionally modified or extended components in the FRA framework. More details on this are provided in individual chapters. However, a brief reference to two modifications is useful at this stage.

Irzik and Nola (2011a, 2011b) initially used the term 'activities' to refer to ideas involving processes used in scientific inquiry. In later work (Irzik & Nola, 2014), they referred to them as "scientific processes". For reasons detailed in Chap. 4, the terms 'activities' and 'processes' are substituted with 'practices'. Using "scientific practices" in the context of the FRA establishes a healthy distance from the over-use and narrow meanings often associated with science process skills in science education, and the generally all-encompassing sense implied by scientific activities. More importantly, it aligns the range of activities involved in this category with those included in the contemporary science education literature (Duschl, Schweingruber, & Shouse, 2007; NRC, 2012).

The original FRA framework (Irzik & Nola, 2011a) included four main categories focused on the cognitive aspects of science. In a revised account, Irzik and Nola (2011b) introduced institutional and social norms as a fifth component that encompassed Merton's norms, social values and research ethics. In a more recent account, the authors (Irzik & Nola, 2014) elaborated on the fifth component by transforming it into a social-institutional dimension. This dimension includes four clearly defined categories: professional activities, scientific ethos, social certification and dissemination, and social values. The authors explicitly give examples of potential categories that can be included but they chose to limit their discussion to four that are non-controversial in nature. Chapter 7 provides a rationale for why additional categories that might be considered by some as controversial (e.g. the economic and colonial aspects of science) should be included under the social-institutional dimension and provides examples for how these categories might be taught in the science classroom.

A final organizational distinction is that the sequence of discussion in Irzik and Nola's (2014) version of FRA is as illustrated in Table 2.1. In other words, they begin the articulation of the FRA with reference to processes of inquiry followed by aims and values, and so on. We deemed it more appropriate to start the articulation and extension of the framework by focusing on the aims and values of science. Focusing on the goals, the targets and embedded values in science should set the pretext for how the subsequent aspects such as practices, methods, knowledge and social-institutional contexts are framed. Although this is an organizational distinction, it also has implications for how the application of FRA in science education can be framed such that its various components make sense particularly from a developmental and cognitive point of view. It would be inconceivable for science students to comprehend and appreciate the value of scientific knowledge without a foundational sense of what science is trying to achieve and how. Likewise the sequence of practices, methods and knowledge also is intended to facilitate the understanding of science in a coherent way.

## 2.4 The FRA as a Holistic Model

How do the components of science as a cognitive-epistemic system relate to those of science as a social-institutional system? This relationship is considered in terms of the graphic representation or model presented in Fig. 2.1 which includes a set of categories that we have added to the Irzik and Nola's (2014) version. The idea can be characterized in the following way. Science as a cognitive-epistemic system occupies a space divided into four quadrants that accommodate its four categories as discussed earlier. This circle floats within a larger concentric one also divided into four quadrants, pertaining to the four components of science as a social-institutional system. The boundaries between the two circles (or spaces) and their individual compartments are porous, allowing fluid movement across. In reality, these

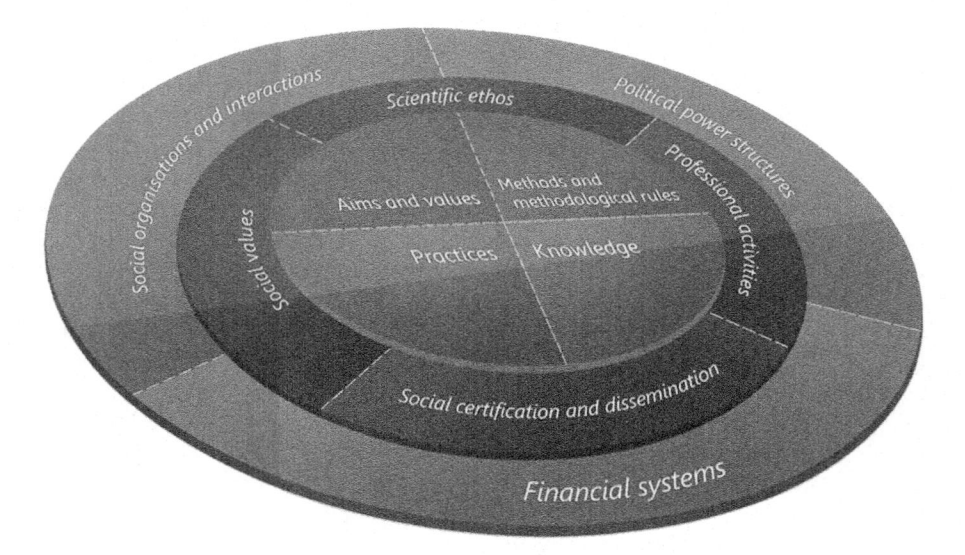

**Fig. 2.1** FRA wheel: science as a cognitive-epistemic and social-institutional system

## 2.4 The FRA as a Holistic Model

components are not compartmentalized but flow naturally in all directions. The purpose of this representation is to provide a visual tool for showing, at-a-glance, how all the components of the cognitive-epistemic and social-institutional systems interact with one another, enhancing or influencing scientific activity. The significance of visualization for facilitating teaching and learning of science is well established (e.g. Gilbert, 2005).

The transformation of the Irzik and Nola's (2014) FRA conceptualization from a textual format to a concentric circle model enhances the depiction of science as a holistic, dynamic, interactive and comprehensive system subject to various influences. Although our representation has to create divisions so as to illustrate the various components, the notion that all of the cognitive, epistemic and social-institutional components co-exist as a whole provides a departure from representing science relative to particular discrete set of ideas. In our view, the image provides a distinctive contribution to research on nature of science (NOS) by offering an interactive, visual and holistic account. These aspects of the representation (and indeed the representation itself) are deemed as improvements to the consensus NOS and FOS frameworks discussed earlier given that their depictions of NOS tend to focus on specific propositions that do not capture adequately the desired degree of breadth and interconnectedness of ideas about science in educational contexts.

In adapting the FRA for science education purposes, we recognized that the social-institutional aspects are limited in Irzik and Nola's (2014) framework. For instance, the political aspects of science were not explicitly acknowledged. Hence we have extended this dimension of FRA to include three additional categories that are discussed in more detail in Chap. 7. We refer to these extra categories as "social organizations and interactions", "political power structures" and "financial systems". The original FRA model has thus been modified to include the additional social-institutional categories as re-represented in Fig. 2.1 by adding the outer-most circle. The reworked framework provides a comprehensive representation of different aspects that characterize the scientific enterprise. Weaving a broader set of social-institutional aspects into the cognitive-epistemic aspects of science is likely to serve a wider range of learners especially those who may not be drawn to the cognitive aspects that dominate school science. The framework serves the agenda of promoting a more balanced and comprehensive account of NOS for all science learners.

Having reviewed the key features of the FRA framework, its adaptation and extension, next we present an example that illustrates how the FRA can be situated in a concrete context. The discovery of the structure of DNA illustrates the broad categories that underlie the FRA framework. James Watson and Francis Crick published the double helix model of DNA in Nature in 1953 (Olby, 1994). Their account was based on the X-ray diffraction image generated by Rosalind Franklin and Raymond Gosling a year earlier as well as information from Erwin Chargaff on the pairing of bases in DNA. Maurice Wilkins and his colleagues had also published results based on X-ray patterns of DNA which provided evidence for the double helix model proposed by Watson and Crick. Watson, Crick and Wilkins were acknowledged jointly for the discovery of the structure of DNA following the death

**Table 2.3** Application of FRA categories to the context of DNA discovery

| FRA | DNA example |
| --- | --- |
| Aims and values | Although the base, sugar and phosphate unit within the DNA was known prior to the modeling carried out by Watson and Crick, the correct structure of DNA was not known. Their quest in establishing the structure of DNA relied on the use of such existing data objectively and accurately to generate a model for the structure. Hence the values exercised included objectivity and accuracy |
| Practices | In their 1953 paper in Nature, Watson and Crick provide an illustration of the model of DNA as a drawing. Hence they engaged in providing representations of the model that they built. They also included the original X-ray diffraction image generated by Franklin on which their observations were based. The scientific practices of representation and observation were thus used |
| Methodology | The methods that Watson and Crick used Franklin's X-ray diffraction data which relied on non-manipulative observation. Hence the methodology involved particular techniques such as X-ray crystallography and observations |
| Knowledge | The main contribution in this episode of science is that a model of the structure of DNA as a double helix was generated. This model became part of scientific knowledge on DNA and contributed to a wide range of scientific disciplines including chemistry, molecular biology and biochemistry |
| Social and institutional context | This episode illustrates some of the gender and power relations that can exist between scientists. There is widespread acknowledgment in the literature and also by Crick himself, for instance, that Franklin was subjected to sexism, and that there was institutional sexism at King's College London where Franklin worked (Sayre, 2000/1975, p. 97). The DNA case also illustrates that science is both a cooperative and a competitive enterprise. Without Franklin's X-rays, Watson and Crick would not be able to discover the correct structure of DNA. This is the cooperative aspect. However there was also competition within and across teams of researchers |

of Franklin. The extent to which Franklin's contribution has been acknowledged has emerged as a contentious issue. In particular, there is widespread recognition that Franklin experienced sexism (Sayre, 2000/1975) (Table 2.3).

The DNA example illustrates how the FRA framework can be applied in science education. Clearly the argument for the inclusion of these various features of science is not new. Numerous science education researchers have already made this argument as is pointed out in the following sections. However, what is novel about this approach is that when covered together, in a collective and inclusive manner, nature of science is presented to learners in a more authentic and coherent fashion. When students confront this and other examples positioned in a similar way (where now comparative aspects across examples can be pursued as well), the "family resemblance" element can also be drawn in. For instance, the precise nature of observation in terms of it being a "scientific practice" in the DNA example can be contrasted with another instance, say, an example from astronomy to draw out the similarities and differences between observation practices in different branches of science.

Identifying the components of science as a cognitive-epistemic and social-institutional system is a beginning step in the design of curricula and lesson materials. We are cognizant of the fact that this example only serves to identify particular topics through which lesson contexts can be generated. The pedagogical strategies that accompany the realization of the FRA framework need to also be considered. Some instructional issues are discussed in Chap. 8 after the components of the system are covered across the book in more detail. There are implications for teacher education as well, in terms of familiarizing science teachers with the content of topics that are likely to be taught in a decontextualised fashion. Teacher educators will need to extend the framework for professional development purposes to support teachers' incorporation of FRA components in their science lessons.

## 2.5 The Relationship of FRA to Research Traditions and Policy in Science Education

It is worthwhile at this stage to discuss how FRA relates to existing research traditions within science education as well as to curriculum policy. The intention is to be illustrative in order to provide a rationale for the relevance of FRA in science education research and policy. In the rest of the book, each component of FRA is covered in more detail in each chapter and more specific links will be made to research and policy.

The FRA framework is related to a wide range of research in science education, which may have historically developed in an unrelated and disparate fashion. The holistic and inclusive nature of the FRA framework opens up opportunities to incorporate for instance, history of science, as well as cognitive models for scientific reasoning, into the design and evaluation of curriculum units. Those opportunities are enhanced by a strong research-base in science education. For example, there is considerable research on students' ideas about the nature of science. Some studies focus on articulating developmental differences in children's understanding of the nature of science (Driver, Leach, Millar, & Scott, 1996; Hammer & Elby, 2000) while other studies document some of the difficulties and successes students encounter with understanding the NOS consensus view (e.g. Lederman, 2007). There is also a plethora of assessment instruments that provide good starting points for developing new formative and summative assessments using findings learned from the application of the VNOSS (Abd-El-Khalick & Lederman, 2000; Lederman, Abd-El-Khalick, Bell, & Schwartz, 2002) and the KNOWS (Allchin, 2012). The literature on socio-scientific issues can inform how investigations of socio-scientific issues contribute to an improved understanding of NOS (Eastwood et al., 2012; Sadler, 2011; Zeidler, Walker, Ackett, & Simmons, 2002). Case studies on NOS implementation from different countries, as well as insights from theoretical studies, can provide useful ideas for developing innovative NOS resources (Grandy & Duschl, 2008; Matthews, 2014). A variety of linguistic and discourse tools can facilitate the implementation of scientific practices (Erduran, 2007; Kelly, 2011; Sandoval, 2005). Curriculum studies can enhance re-conceptualizing the integration of integrating an FRA approach to NOS teaching (Donnelly, 2001; Rudolph, 2000; Schwab, 1964). Finally, studies on the critical use of history of science (Allchin, 2013; Erduran, 2001; Matthews, 1994, 2012; Milne, 1998) can be used to enrich instruction on nature of science.

In addition to its compatibility with these research traditions, the FRA is also compatible with policy frameworks such as past (AAAS, 1989; NRC, 1996) and recent science education reforms in the USA (NRC, 2012). Even though the *Framework for K-12 Science Education [FKSE]* (NRC, 2012) does not designate a specific chapter to discuss the nature of science as the *Science for All Americans [SFAA]* document did, the spirit of NOS is integrated throughout its content. The *FKSE* calls for a triadic emphasis on three dimensions: scientific and engineering practices, disciplinary core ideas, and crosscutting concepts. These dimensions are expected to be taught in an interrelated and coherent way leading to the realization of a normative goal in which "students should develop an understanding of the enterprise of science as a whole—the wondering, investigating, questioning, data collecting and analyzing" (NGSS Lead States, 2013, p. 1). This meta-level of understanding aligns well with the categories of the FRA. In Table 2.4, we list a few examples of how categories of the FRA correspond to the vision promoted in the *Framework for K-12 Science Education* (2012) and to expectations about students' understanding of the nature of science based on Appendix H in the *Next Generation Science Standards* (NGSS Lead States, 2013). These examples are not the only ones that can be found in the documents, but they represent well the ideas contained therein. Even though the reform vision and ensuing standards may not be directly relevant to readers outside the United States, we believe that a similar analytical process can be undertaken with curriculum standards of other countries.

Although there seems to be some overlap of the FRA categories with existing statements in policy recommendations, the particular ways in which policy statements articulate, or fail to articulate, aspects of the FRA becomes an issue. For instance, take the reference to the "Social and Institutional Context" category from Table 2.4. The statements are rather broad and do not necessarily indicate which aspects of the social or the institutional dimensions of science are to be emphasized and how. It is also not clear where such dimensions need to be included in science lessons. If the emphasis is on cognitive-epistemic and social-institutional contexts becomes an add on, the goal of presenting science to learners in a holistic fashion is lost. What results is that the various dimensions of science are emphasized and prioritized selectively and persistently while others become peripheral and 'cosmetic' to serve a very generic and broad goal. The outcome of such an approach is that students learn a distorted, decontextualized and incoherent view of the nature of science.

## 2.6 Potential Challenges in Applying the FRA in Science Education

**Table 2.4** Alignment of FRA categories with recent reform documents in the USA

| FRA | Framework for K-12 Science Education (NRC, 2012) | Next Generation Science Standards (NGSS Lead States, 2013) |
|---|---|---|
| Aims and values | "Epistemic knowledge is knowledge of the constructs and values that are intrinsic to science." (NRC, 2012, p. 79) | "Science Addresses Questions About the Natural and Material World." |
| | | "Scientific information is based on empirical evidence." (p. 4) |
| Practices | "…important practices, such as modeling, developing explanations, and engaging in critique and evaluation (argumentation)… Engaging in argumentation from evidence understanding of the reasons and empirical evidence for that explanation, demonstrating the idea that science is a body of knowledge rooted in evidence. (p. 44) | "Students must have the opportunity to stand back and reflect on how the practices contribute to the accumulation of scientific knowledge…. Through this kind of reflection they can come to understand the importance of each practice and develop a nuanced appreciation of the nature of science." (p. 7) |
| Methodology | "Practicing scientists employ a broad spectrum of methods…" (NRC, 2012, p. 44) | "Scientific Investigations Use a Variety of Methods." (p. 4) |
| Knowledge | "Students need to understand what is meant, for example, by an observation, a hypothesis, a model, a theory, or a claim and be able to distinguish among them." (NRC, 2012, p. 79) | "Science is a Way of Knowing." |
| | | "Scientific Knowledge is Open to Revision in Light of New Evidence." |
| | | "Scientific Models, Laws, Mechanisms, and Theories Explain Natural Phenomena." (p. 4) |
| Social and institutional context | "Seeing science as a set of practices shows that theory development, reasoning, and testing are components of a larger ensemble of activities that includes networks of participants and institutions…." (p. 43) | "Science is a Human Endeavor" (p. 4) |

## 2.6 Potential Challenges in Applying the FRA in Science Education

The brief description of the FRA categories in this chapter may perplex the reader on different levels. For starters, the approach seems complex. It groups NOS ideas in unfamiliar ways; seems to place high cognitive demands on students; and may seem challenging to teachers. This section addresses some of these potential concerns.

The apparent complexity of the FRA is precisely its core strength. It is complex at first sight, yet it is simple in terms of helping organize thinking about a large number of pedagogically appropriate NOS ideas in terms of few inter-related categories. Because it is not prescriptive at the level of specifying curriculum and instructional actions, the FRA leaves educators with a wide range of choices regarding how to embed some of these ideas from each of the five categories in their teaching. This range of choices is advantageous because it does not mandate a specific set of ideas to be taught in relation to a given content, but invites the selection of relevant ideas along each category as they relate to the content. Educators seeking a short list of NOS statements to incorporate into classroom instruction will find instead guiding principles that need to be unpacked and embedded within the content they are teaching. These guiding principles are not declarative statements. They are contextual domains (cognitive, epistemic, social and institutional) that can be explored and translated into practical teaching and learning outcomes.

As for familiarity, the FRA deals with some commonly discussed themes in the science education literature, such as scientific practices, scientific methodology, and social certification. Some of the categories we introduced may seem either marginal or controversial to bring to students' attention. For example, the financial aspects of science and commodification of scientific knowledge discussed in Chap. 7 might communicate a rather pessimistic image of the scientific enterprise. The pedagogical implications of including or excluding such discussions in the classroom are addressed, but not necessarily settled.

In the end, we believe that more discussion and debate on these issues are needed beyond this book which is the starting, not the end point for a new debate on nature of science. Furthermore, it will be important to improvise effective models for communicating the notion of science as social system in school science especially with regards to how to balance its familiar components (e.g. socio-scientific issues) with less familiar ones (e.g. colonial science). Further research and development of models for incorporating these ideas into the core curriculum, instruction, and professional development will be needed. This is an ambitious task that can incorporate the work of many researchers who passionately believe that it is possible for students and teachers to access these ideas if we design the right curriculum materials and structure the appropriate learning environment to implement them.

It could be argued that applying the FRA to the curriculum might increase the cognitive demands on students and push the content beyond their reach. However, "cognitive development and educational psychology are converging on important conclusions that address policy concerns about STEM illiteracy. All show that we can teach science in a meaningful and better way, much earlier than we have—and that even preschool children have some relevant abstract abilities" (Vandell, Gelman, & Metz, 2010, p. 26). We extend the logic of this argument to maintain that when appropriate epistemic and social aspects are intertwined with the cognitive ones, they provide a stronger context and deeper meaning to the learning experience (Dagher, 2012). When these epistemic components are infused in a developmentally appropriate way, children will most likely understand them. A companion learning

## 2.6 Potential Challenges in Applying the FRA in Science Education

progression for these ideas can be developed in relation to the FRA, but this goes beyond the parameters of the present task.

The pedagogical demands that FRA might place on teachers may seem unreasonable. Teachers would need to know a lot more about how the FRA categories are contextualized for instance in the American context, within scientific practices, cross-cutting concepts and core ideas. Teachers need to have access to additional information, practical resources, and suggestions on how to promote more holistic discussions about nature of science. We acknowledge this to be a normal task that follows the introduction of new frameworks. What the FRA does is help teachers organize how they might draw on existing resources pertaining to each of the categories of the FRA. When internalized, the incorporation of these ideas is expected to flow out of planned inquiries into scientific practices, or discussions on how scientific knowledge is impacted by financial and other socio-cultural factors. Specific probes and supplements to activities can be added that promote the meta-cognitive thinking about these issues. Less important activities can be removed.

The effectiveness of the FRA model is yet to be investigated. The development of the FRA for educational use at this current stage is primarily conceptual and must be followed up with additional translational work that involves curriculum revision followed by empirical studies to determine optimal design of effective science curriculum and instruction. Interventions based on this framework need to be studied in terms of their effectiveness to improve students' understanding of nature of science and of science concepts. Our primary task in this book is to make the case that the expanded FRA can be a fruitful new conceptual territory that can redefine and rejuvenate research on the nature of science in science education. Adaptations of the examples presented throughout the book into empirical research will be crucial in illustrating the practical dimensions of the FRA model.

There are various possible processes and outcomes for how applications of the FRA can be characterized. It could be that we, as science educators, are borrowing from the work of philosophers of science in a way to repeat an existing framework for the purpose of generating a list of ideas for inclusion in science education. This sense of the application is about repetition of existing ideas for educational purposes. The primary outcome of this approach would be the generation of a list of concepts that are deemed to be useful for science education. A second approach could be translation of philosophical perspectives for use in science education. This sense of 'translation' would still yield a list as an outcome. However the list would be pedagogically mindful of how the philosophers' account maps to education, and it would be an applied list. A third sense of application concerns expansion of the philosophical work to have an original contribution. Here, the main outcome would be an extended list with new content. A fourth sense would involve the extension and translation where the now extended list is mapped to its pedagogical purposes.

A final sense of the way that philosophical analysis can be used for science education purposes concerns not just an extension and a translation of a set of original ideas but rather a complete transformation of a germ of an idea guided by pedagogical purposes where the key outcome now constitutes an original synthesis. It is in this final sense of the application of FRA to science education that we consider our work to be situated. In using, extending and transforming the original FRA, we are producing a new framework that has a different purpose and content as well as potential to redefine nature of science for science education. The original FRA is now reconfigured to project an image of science that is holistic but not normative in what it promotes for science teaching and learning. This image is not stagnant but is generative and malleable in nature, giving rise to multiple possibilities. The primary contribution of this approach is that the outcome of the application produces a set of heuristics that are not only epistemologically sound but are also pedagogically relevant and meaningful.

In summary, we propose the FRA as a practical conceptual tool to organize the infusion of various aspects of nature of science into the curriculum. Some of the ideas in each of the categories may apply to some science content, while others may apply better to other content. So while it is optimal that as many categories be addressed as possible when exploring a scientific unit of study, it is not necessary that the same level of depth be achieved for all components. It is to be expected that some will be addressed more than others on different occasions, but that over the school year or across grade levels, all aspects would have been addressed meaningfully and in context. Selecting and packaging FRA components to achieve specific NOS goals must be coordinated with other science education goals and with developmentally appropriate NOS content.

## 2.7 The Layout of the Book

In the rest of the book, a chapter is devoted to the discussion of each of the four categories under science as a cognitive-epistemic system, and one chapter for discussing the 11 categories under science as a social-institutional system. The discussions in each chapter are supplemented by instructional examples. In Chap. 3, we focus on aims and values and their role in science and emphasize their cognitive and epistemic aspects. In the discussion, following questions are explored: What are the aims and values of science? How do they guide scientific practices and theory choice? How do values influence the growth of scientific knowledge? Aims and values of science from various philosophical viewpoints are discussed and implications for science education are drawn. Furthermore, specific examples are drawn to demonstrate how scientific aims and values can be promoted in science lessons.

We discuss the range of scientific practices that scientists use in Chap. 4 where the following questions are addressed: What are the key epistemic, cognitive and social practices of science? How are these practices generated, evaluated and revised? The discussion is centered on three examples of scientific activities, namely classification, observation and experimentation. The choice of these activities rests on their prevalence in some version within the international science curricula. After reviewing select aspects of the nature of these activities, we illustrate how reflection on these scientific activities can be envisaged as part of a comprehensive model of

## 2.7 The Layout of the Book

scientific practices that would ensure that they are not visited in a fragmented fashion in science classrooms. A visual tool of scientific practices is proposed that consolidates some existing and contemporary accounts from curricular policy documents with implications for science curriculum and instruction.

After raising issues about the different ways by which the scientific method has been defined, Chap. 5 focuses on scientific methods and methodological rules. The question of what methods are best suited for investigating scientific problems in different domains is raised, and a pedagogical framework for communicating a range of scientific methods used in different science sub-disciplines is presented. A set of pedagogical strategies are proposed that can be used for promoting a concrete contextual understanding of the diversity of scientific methods. This chapter is particularly important in its clear depiction of the diversity of methods used in science which sits in contrast to the often over-emphasized and caricaturized image of the scientific method.

In Chap. 6, forms of scientific knowledge that include laws and models are described. The discussion is guided by the following questions: What are the different products of science? How are these forms of scientific knowledge related? How are they produced? What function/role do they play in the development of knowledge claims? Are there disciplinary variations in theories, laws, and models? What is the relationship of explanation to theories, models, and laws? Why is it useful for students to understand various forms of scientific knowledge? The chapter concludes by discussing ways for promoting discussions on the growth of scientific knowledge more systematically in educational contexts. Although school science is cluttered with scientific knowledge, often the processes of knowledge growth are not effectively articulated at the level of the classroom. As a result, students do not develop a sense of how scientific knowledge is generated, evaluated and revised throughout its development. Establishing some models of growth of scientific knowledge that can be effectively used in science lessons can help facilitate students' meaningful understanding of scientific knowledge.

Focusing on the four original FRA categories of science as a social-institutional system in Chap. 7, this dimension is extended to include three additional categories. After describing the system's components, we discuss a range of additional social conceptions of science that are not traditionally highlighted in school science. The following questions are addressed: What political, economical and sociological factors drive the scientific enterprise? How are scientists and communities of scientists influenced by such factors? The main purpose of this chapter, then, is to outline a set of social and institutional contexts that illustrate the scientific enterprise. Often in school science, the organizational and institutional aspects of science are particularly missing. For example, how scientists work in groups, the organizational and financial dynamics that govern scientists' behaviors and decision-making are not themes that are regularly captured in science lessons.

In Chap. 8, we revisit the FRA and its categories and how they work synergistically to provide a holistic account of science. The following questions are raised: What pedagogical strategies would go with which type of goal in these examples? How can teachers be supported in the development of their understanding and implementation of such holistic accounts of science? We also illustrate how using the FRA framework brings coherence to the science curriculum as it allows the adoption of effective teaching strategies based on decades of science education research. The connections between the FRA approach and the *Next Generation Science Standards* (NGSS Lead States, 2013) are explored given the timeliness of this document. Considering the impact of previous curriculum reform documents from the United States in the rest of the world, for instance the 1996 National Research Council published *National Science Education Standards*, it is likely that NGSS will gain much attention worldwide beyond the publication timeline of this book. Hence the intention is to offer some insight to the international science education research and policy audience regarding how our approach maps onto emerging curricular goals. The chapter concludes with a set of implications for an empirical research agenda.

We conclude this chapter with a word of caution. Irzik and Nola's (2014) version of the FRA includes eight-categories, and our extension leads to 11. The suggestion is not a replacement of an existing NOS "consensus view" that practically relies on a set of seven tenets, for instance, with a set of 11 categories. The approach in the application of FRA is more nuanced in the following way. First, the adaptation of the FRA is made with appeal to theoretical arguments on 'science' based on contemporary research philosophy of science. Second, the transformation of FRA principles to science education practice is based on our understanding of cognitive science and science education research which have provided a solid knowledge base of what students and teachers know and are capable of doing. We also base it on our collective experience (four decades), in the field and keen awareness of exemplary teaching practices. Third, rather than listing a set of NOS learning objectives focusing on a limited set of ideas, overarching principles are outlined from which objectives can be drawn and adapted to different settings and grade levels. The overarching principles invite teachers and teacher educators to be creative participants in seizing opportunities for discussing the nature of science, in context, along the 11 categories highlighted in this book.

## References

AAAS. (1989). *Science for all Americans*. Washington, DC: American Association for the Advancement of Science.
Abd-El-Khalick, F., & Lederman, N. (2000). Improving science teachers' conceptions of nature of science: A critical review of the literature. *International Journal of Science Education, 22*, 665–701.
Allchin, D. (2012). Toward clarity on whole science and KNOWS. *Science Education, 96*(4), 693–700.
Allchin, D. (2013). *Teaching the nature of science: Perspectives and resources*. St. Paul, MN: SHiPs.
Dagher, Z. (2012, March 25–28). *Re-imagining nature of science: Implications for policy and research*. Paper presented at the annual meeting of the National Association for Research in Science Teaching, Indianapolis, IN.

# References

Donnelly, J. (2001). Contested terrain or unified project? 'The nature of science' in the national curriculum for England and Wales. *International Journal of Science Education, 23*, 181–195.

Driver, R., Leach, J., Millar, R., & Scott, P. (1996). *Young people's images of science*. Buckingham, UK: Open University Press.

Duschl, R., Schweingruber, H., & Shouse, A. (Eds.). (2007). *Taking science to school*. Washington, DC: National Academies.

Eastwood, J. L., Sadler, T. D., Zeidler, D. L., Lewis, A., Amiri, L., & Applebaum, S. (2012). Contextualizing nature of science instruction in socioscientific issues. *International Journal of Science Education, 34*, 2289–2315.

Erduran, S. (2001). Philosophy of chemistry: An emerging field with implications for chemistry education. *Science & Education, 10*(6), 581–593.

Erduran, S. (2007). Breaking the law: Promoting domain-specificity in chemical education in the context of arguing about the Periodic Law. *Foundations of Chemistry, 9*(3), 247–263.

Gilbert, J. (Ed.). (2005). *Visualisation in science education*. Dordrecht, The Netherlands: Springer.

Grandy, R., & Duschl, R. (2008). Consensus: Expanding the scientific method and school science. In R. Duschl & R. Grandy (Eds.), *Teaching scientific inquiry: Recommendations for research and implementation* (pp. 304–325). Rotterdam, The Netherlands: Sense Publishers.

Hammer, D., & Elby, A. (2000). Epistemological resources. In B. Fishman & S. O'Connor-Divelbiss (Eds.), *Fourth international conference of the learning sciences* (pp. 4–5). Mahwah, NJ: Erlbaum.

Irzik, G., & Nola, R. (2011a). A family resemblance approach to the nature of science. *Science & Education, 20*, 591–607.

Irzik, G. & Nola, R. (2011b). *A family resemblance approach*. Plenary presentation session with N. Lederman titled: Current philosophical and educational issues in nature of science (NOS) research, and possible future directions. Presented at the International History, Philosophy and Science Teaching Conference, Thessaloniki, Greece.

Irzik, G., & Nola, R. (2014). New directions for nature of science research. In M. Matthews (Ed.), *International handbook of research in history, philosophy and science teaching* (pp. 999–1021). Dordrecht, The Netherlands: Springer.

Kelly, G. J. (2011). Scientific literacy, discourse, and epistemic practices. In C. Linder, L. Östman, D. A. Roberts, P. Wickman, G. Erikson, & A. McKinnon (Eds.), *Exploring the landscape of scientific literacy* (pp. 61–73). New York: Routledge.

Lederman, N. (2007). Nature of science: Past, present, future. In S. Abell & N. Lederman (Eds.), *Handbook of research on science education* (pp. 831–879). Mahwah, NJ: Lawrence Erlbaum.

Lederman, N., Abd-El-Khalick, F., Bell, R., & Schwartz, R. (2002). Views of nature of science questionnaire: Toward valid and meaningful assessment of learners' conceptions of nature of science. *Journal of Research in Science Teaching, 39*, 497–521.

Matthews, M. (1994). *Science teaching: The role of history and philosophy of science*. New York: Routledge.

Matthews, M. (2012). Changing the focus: From nature of science (NOS) to features of science (FOS). In M. S. Khine (Ed.), *Advances in nature of science research* (pp. 3–26). Dordrecht, The Netherlands: Springer.

Matthews, M. (Ed.). (2014). *International handbook of research in history and philosophy for science and mathematics education*. Dordrecht, The Netherlands: Springer.

Milne, C. (1998). Philosophically correct science stories? Examining the implications of heroic science stories for school science. *Journal of Research in Science Teaching, 35*, 175–187.

National Research Council. (1996). *National science education standards*. Washington, DC: National Academies Press.

National Research Council. (2012). *A framework for K-12 science education: Practices, crosscutting concepts, and core ideas*. Washington, DC: National Academies Press.

NGSS Lead States. (2013). *Next generation science standards: For states, by states*. Appendix H. Retrieved from http://www.nextgenscience.org/next-generation-science-standards

Olby, R. C. (1994). *The path to the double helix: The discovery of DNA*. New York: Dover Publications.
Resnik, D. (2007). *The price of truth*. Oxford, UK: New York.
Rudolph, J. (2000). Reconsidering the 'nature of science' as a curriculum component. *Journal of Curriculum Studies, 32*, 403–419.
Sadler, T. D. (Ed.). (2011). *Socio-scientific issues in the classroom: Teaching, learning and research*. New York: Springer.
Sandoval, W. (2005). Understanding students' practical epistemologies and their influence on learning through inquiry. *Science Education, 89*, 634–656.
Sayre, A. (2000/1975). *Rosalind Franklin and DNA*. New York: W.W. Norton & Co.
Schwab, J. J. (1964). The structure of the disciplines: Meaning and significances. In G. W. Ford & L. Pugno (Eds.), *The structure of knowledge and the curriculum*. Chicago: Rand McNally.
Vandell, D., Gelman, R., & Metz, K. (2010). Early learning in science. In National Academy of Engineering, *STEM Summit 2010: Early childhood through higher education*. Retrieved from http://ocstem.org/files/STEMSummit2010Report.pdf
Zeidler, D. L., Walker, K. A., Ackett, W. A., & Simmons, M. L. (2002). Tangled up in views: Beliefs in the nature of science and responses to socioscientific dilemmas. *Science Education, 86*(3), 343–367.

# Chapter 3
# Aims and Values of Science

The chapter explores the role of aims and values in science. In particular, the epistemic and cognitive aims and values are emphasized, as the social, political and cultural aims and values, are revisited in Chap. 7 in the discussion on social contexts of science. To guide the discussion in this chapter, the following example questions are posed: What are the aims and values of science, and how do values function? For instance, what values come into play when scientists choose between theories? Do values apply similarly across different functions in science? How do values limit or expand scientific knowledge? The components of scientific aims and values as described by various philosophers of science are discussed, and the review is extended to draw some implications for science education. Examples are drawn to show how scientific aims and values can be promoted in science lessons particularly in relation to assessment of a range of values.

## 3.1 Introduction

Values in relation to science can be considered from epistemic, cognitive, cultural, social, political, moral and ethical perspectives. Considering the vast amount of work within philosophy of science and sociology of science on the subject of values, and the ongoing debate about their distinction, the aim is to clarify and summarize a functional notion of values for science education purposes. In this chapter, the focus is on epistemic and cognitive values in science, and in Chap. 7 when the broader context of science is revisited, the discussion is extended to social norms and cultural values. Such separation is artificial in nature and is only intended for ease in coverage of a wealth of concepts. Values and norms in science are interrelated features of science that are difficult to disentangle. In this sense, we are in agreement with Longino (1995) and Allchin (1999) that heterogeneity of values is a resource to scientific objectivity, not a weakness. We are also in agreement with Carrier (2013) who argues for epistemic pluralism in relation to scientific values. However, we should caution, as do Allchin and Longino, that subscribing to heterogeneity and pluralism in values in science and their import in science education does not necessarily imply a relativist position. To the contrary, we believe that it is the diversity of values that ensures the "check and balance" of scientific claims and enables the generation of robust scientific knowledge.

In their 2011 *Science & Education* paper, Irzik and Nola review the aims and values of science in the following way:

> The aims in question are not moral but cognitive. Of course, there are many other aims in science such as consistency, simplicity, fruitfulness and broad scope (Kuhn, 1977); high confirmation, as emphasized by logical empiricists (Hempel, 1965, Part I); falsifiability and truth or at least verisimilitude (i.e. closeness to truth) (Popper, 1963, 1975); empirical adequacy (van Frassen, 1980), viability (von Glasersfeld, 1989), ontological heterogeneity and complexity, as emphasized by empiricist feminists like Longino (1997). (p. 597)

Allchin (1999), on the other hand, distinguishes between epistemic and cultural values. Epistemic values are those that "guide the pursuit and methods of science", while cultural values enter science through the work of individual scientists (p. 1). While subsequently Irzik and Nola (2014) classify aims and values of science under science as a cognitive and epistemic system, Allchin talks about ultimate and proximate values. Among the proximate set of values, he distinguishes between institutional imperatives (Merton's), epistemic values, and social values. By epistemic values he refers to controlled observation, interventive experiments, confirmation of predictions, repeatability, and statistical analysis. Like Longino (1995), Allchin argues that diversity of values enables the production of more robust knowledge, and that is precisely such diversity and the communal justification of knowledge claims that science exercises objectivity. According to Allchin, science and values intersect in at least three ways. First, there are epistemic values that guide research. For example, accuracy, testability and novelty can guide scientists in making judgments about knowledge claims. Second, science is situated in a particular cultural context and thus it is inherently composed of and surrounded by values that are practised by scientists. Some of the examples provided by Allchin include the role of gender and race in rationalization of scientific knowledge. Third, science itself can generate values that can contribute to society, culture and ethics. For example, science as a problem-solving activity is an aspect of science that can have societal uptake that can result in influencing other social systems.

Beyond a basic characterization of aims and values of science, Irzik and Nola (2011) illustrate how different philosophical perspectives would differ on their take of the aims and values of science. For example, they highlight the scientific realist and anti-realist positions on observation. They also illustrate how some philosophers' accounts might be in direct opposition to others', like Longino's "ontological heterogeneity" and 'complexity' clashing with Kuhn's 'simplicity' as aims of science. They resolve some of these tensions by appealing to the Family Resemblance

3.2 What Are Aims and Values of Science? 43

Approach through a nuanced discussion illustrating the complexity in representing different philosophical stances on scientific aims and values:

> …a scientific realist interpretation of scientific theories will share truth at the observable level as an aim of science with a constructivist empiricist or a Kuhnian interpretation, but differ from them at the unobservable level; similarly, scientific realists and constructivist empiricists will agree that explanation is an aim of science, but Duhemians will disagree; and so on. In this way, we will have a family resemblance with respect to the aims of science according to different philosophical interpretations or stances. (Irzik & Nola, 2011, p. 598)

Irzik and Nola's present a nuanced approach to the positioning of different philosophical approaches on scientific aims and values. For example, when they discuss the values of 'simplicity' and 'explanatoriness' they illustrate how these values can serve different functions depending on the way they are used, such as evaluation criteria for theory choice, and components of methodological rules in science:

> It is also to be noted that values in science also function as criteria for theory choice and can be expressed as methodological rules. Take, for example, the value of simplicity. We can write this as the rule: given two rival theories, other things being equal, choose the simpler theory. Similarly, the value explanatoriness gives the following criterion for theory choice: given two rival theories, other things being equal choose the theory that is more explanatory. (Irzik & Nola, 2011, p. 598)

In their chapter in the *Handbook on History, Philosophy and Science Teaching* edited by Michael Matthews, the authors revisit the theme of aims and values (Irzik & Nola, 2014). They add to their previous list of aims the notions of 'prediction' and 'explanation' and note that these two aims are typically neglected by the science education literature. Irzik and Nola argue that scientists tend to value not just any predictions but novel predictions. The rest of the chapter will unpack the aim of prediction and refer to the work of other philosophers of science (e.g. Douglas, 2000). In terms of the aim of 'explanation', they point out that "all scientific explanations are naturalistic in the sense that natural phenomena are explained in terms of other natural phenomena, without appealing to any supernatural or occult powers and entities" (Irzik & Nola, 2014, p. 1004). The reference to Irzik and Nola's work provides not only an overview of aims and values of science but also highlights example tensions and nuances related to them. Both of these instances (i.e. a set of aims and values, and particular tensions and nuances) could potentially provide some suggestions for science education. These suggestions are reviewed in subsequent sections after questioning further characterizations of aims and values in science.

## 3.2 What Are Aims and Values of Science?

As Irzik and Nola (2011) indicated, 'aims' and 'values' are at times difficult to distinguish because they might serve similar functions. In approaching the aims and values of science, we sought to address a broad range of issues that might be applicable in different disciplines and educational contexts.

Aims and values of science have centered in the long-standing debates on objectivity in philosophy of science. The notion of neutrality of scientific claims as devoid of bias and individual subjective prejudice have led to the dichotomy of objectivity and subjectivity, and the separation of scientific fact from subjective interpretation. The earlier depictions of objectivity were grounded in individually-centered accounts (e.g. Francis Bacon's work) where no significance was placed on interactions among scientists. Subjectivity was based on individual psychological bias and prejudice that interfered with objectivity of science. Some philosophers of science acknowledge subjectivity at the level of the individual scientists but claim a level of 'objectivity' of the produced knowledge. Longino's notion of 'intersubjectivity' of science captures this tension between the interplay of scientific facts and values. Harding (1991) also problematizes the notion of objectivity by distinguishing between 'strong' and 'weak' objectivity. More recently, the study of 'bias' in science has received serious attention (Gluud, 2006; Resnick, 2007). According to Wilholt (2009), while one of the values in science upholds the ideal of neutrality, strict adherence to scientific methods does not eliminate all bias because other values are at play at different phases of inquiry. Bias in scientific research (distinguished from willful falsification of data) is inherent in various facets of the scientific enterprise, ranging from the process of selecting the research question and designing the research methodology, to the communication of findings and dissemination of results (Wilholt, 2009). Wilholt identifies various types of bias, for instance, preference bias and publication bias. For instance, preference bias "occurs when a research result unduly reflects the researchers' preference for it over other possible results" (Wilholt, p. 92).

A recent argument put forward by Carrier (2013) aimed to transcend the difficulties in the dichotomy of objectivity and subjectivity in science in relation to values by appealing to different positioning of epistemic pluralism and consensus formation. According to Carrier, scientific reasoning and scientific community practices can be brought into harmony when epistemic pluralism and consensus formation are considered separately. Among the various epistemic values and attitudes specified by Carrier are "respect for rational argument, commitment to gain objective knowledge and appeal to shared epistemic goals" (p. 2548). The conceptualization of values in this sense introduces the language that is relevant to science educators. The reference to 'respect' and 'commitment' emphasizes the value element involved in epistemic practices of science. In contrast, numerous curricular documents that advocate the acquisition of scientific values in science education have rather broad goals that do not necessarily differentiate between epistemic actions and epistemic attitudes. For example, promoting scientific habits of mind has been a key idea in reform efforts for some time (e.g. AAAS, 1989), focusing on a number of attitudes, values and skills that should be included in science teaching. The AAAS document describes four values and attitudes that include: (a) values inherent in science, mathematics, and technology [SMT] (they encompass aspects of the nature of these disciplines), (b) reinforcement of general societal values (need to foster curiosity, openness to new ideas, and informed skepticism), (c) the social value of SMT (develop critical attitudes towards science), and (d) attitudes towards learning SMT. However the document neither clearly justifies these values nor grounds them in the extant literature. Details on what these categories mean, what they involve, and how they impact science learning are not discussed.

Framing of epistemic values through the work of philosophers of science brings a language that is nuanced and potentially more fruitful in application to science education. For example, the difference between "students should produce rational arguments" and "students should possess the epistemic value of respect for rational arguments" may be subtle but has significant implications for how they occur at the level of the classroom. Whereas the first prioritizes a scientific practice, the latter places a stronger emphasis on epistemic attitudes and values, and the particular lesson resources produced for student use will equally vary with respect to the priority and emphasis in the instructional goals. The philosophical literature clarifies not only the definition and differentiation of epistemic attitudes and values but also provides some indicators for their function and use in specific circumstances in science. For example, Carrier (2013) discusses how epistemic values may come into play in making a choice between empirically equivalent hypotheses:

> If two accounts are empirically equivalent and one of them uses a large number of unrelated hypotheses while the other one appeals to a few overarching principles, the commitment to coherence (or simplicity or broad scope) favors the latter approach. Assessed in light of this value, the evidence favors the more unifying treatment—even if the two approaches are empirically equivalent. The scientific community resorts to such values for making a choice between empirically indistinguishable alternatives. Scientists break the tie between rival accounts that conform to the data to approximately the same degree by appeal to virtues that transcend the requirement of empirical adequacy. (Carrier, 2013, p. 2551)

In this instance, there is an appeal to the value of "commitment to coherence" with overarching principles, where empirical scrutiny may not be adequate to differentiate alternative hypotheses. This statement also recognizes that the resolution of rival accounts can involve appeal to non-cognitive, "non-empirical or superempirical" values that "cannot be based on experience alone" (Carrier, 2013, p. 2555). Longino has pointed out that these non-empirical values can be socio-political (Longino, 1995), and in this sense her analysis shows the intricate relationship between the epistemic and the social aims:

> Empirical adequacy and accuracy (treated as one or separate virtues) need further interpretation to be meaningfully applied in a context of theory choice. Those interpretations are likely to import the socio-political or practical dimensions that the search for a purely cognitive criterion seeks to escape. At the very least the burden of argument falls on those who think such an escape possible. (Longino, 1995, p. 395)

Furthermore, Longino also discusses how the devaluation or exclusion of empirical data, due to the focus on a limited set of factors pertaining to specific observational or experimental conditions, can result in distorted interpretations, thus leading to constrained knowledge about the observed event or the effect. Longino put forward "procedural standards" that are intended to govern the process of critical examination in science. One of her requirements concerns the need to "take up criticism and to respond to objections" appropriately (Longino, 1993,

p. 267, 2002, pp. 129–130). This requirement broadens the Popperian obligation to "address anomalies and counter instances" (Popper, 1957, pp. 66–69). Scientists strive to seriously consider challenges and to deal with them accordingly. This community rule is supposed to preclude personal or institutional power playing; arguments should be appreciated independently of community hierarchies (Longino, 2002).

Longino suggests empirical adequacy, novelty, ontological heterogeneity, complexity of interaction, applicability to human needs, and decentralization of power as standards for assessing scientific theories (Longino, 1995 cited in Carrier, 2013).

> When we detach a factor from the contexts in which it naturally occurs, we are hoping to achieve understanding of that factor's precise contribution to some process. But by taking it out of its natural context we deprive ourselves of understanding how its operation is affected by factors in the context from which it has been removed. This is, of course, a crucial aspect of experimental method. I suspect that it's not (or not always) the decontextualization that is to be deplored, but the concomitant devaluation as unimportant or ephemeral of what remains. (Longino, 1995, p. 395)

Carrier (2013) argues that epistemic significance is determined by epistemic values. As an example, he illustrates that in some cases, large scale generalizations may be deemed epistemically significant depending on the research questions, particularly when a large number of propositions are likely to indicate the truth of an overarching issue. So the value of "favor large scale generalizations" would be suitable when overall theoretical coherence is sought. However a large number of isolated propositions (e.g. identifying the number of leaves of a tree at a given time) will not be considered significant given it is not nested in an overall theoretical framework. Hence, the value around favoring large scale generalizations and their epistemic significance is specified. In other words, this value is not a generally applied value regardless of the research question asked but rather its epistemic significance is tightly related to the research goals.

The discussion of values in science has also generated considerable debate about the ways in which key practices of science such as observation and experimentation are themselves not immune from the impact of values. This issue has attracted a great deal of attention by feminist scholars, among others, who elaborated on the precise nature of observation bias. Note how Longino considers gender in the study of speciation in biology and the notion of a decontextualized variable in experimentation:

> The failure to attend fully to the interactions of the entire social group, including its females, in studying the males of a species has led to distorted accounts of the structure of animal societies, including male-male interactions. In toxicity studies, the focus on a single chemical's toxic properties fails to inform us how its activity is modified, canceled or magnified by interaction with other elements in its natural environments. Focus on gene action has blinded us to the ways in which the genes must be activated by other elements in the cell. These models may well be empirically adequate in relation to data generated in laboratory experiments, but not in relation to potential data excluded by a particular experimental set up. (Longino, 1995, p. 395)

Allchin (1999) on the other hand emphasizes the role of Giere's cognitive resources in shaping scientific knowledge. Cognitive resources include not only the

## 3.2 What Are Aims and Values of Science?

concepts, interpretive frameworks and motivations but also the values that are based on experience. These values play a part in shaping the formation of new knowledge because they serve as a filter for determining new research questions, research design and interpreting results. Using the case of craneologists, Allchin discusses how scientists were adamant about concluding from myriad measurements they conducted that women were inferior to men, until two women at the turn of the century provided evidence to the contrary and confirmed that the margin of error in the previous studies was larger than the noted relationship. Allchin maintains that having diverse values in scientific practice provides a self-correcting mechanism. The more diverse the scientists (and the cognitive resources they bring along) involved in solving a given problem, the less likely the results will be biased. This is so because while "identifying and remedying error takes work" (Allchin, 1999, p. 6) and is not automatic, it is largely dependent on the epistemic value of "criticism and responsibly addressing criticism" (Allchin, 1999, p. 6) that characterize scientific practice.

The difficulty in disentangling values as either epistemic or cultural and social can be illustrated by an example proposed by Machamer and Douglas (1999). The example illustrates social values in the context of industry; social interactions between scientists and their employers; and the values of social utility and respect for human life. The authors discuss a case study on the epidemiology of dioxins. Dioxins are chemicals that are by-products of many industrial processes, and they are very toxic in small doses. The danger of dioxins to human health is not contested, although the amount of how much is toxic has been debated for many years. Experimentation with animals in the laboratory has indicated that dioxins can lead to birth defects, cancer and reduced immunological response. In the United States, the National Institute for Occupational Safety and Health initiated a dioxin registry to trace workers who were exposed to dioxin-contaminated herbicides (Fingerhut et al., 1991). The resulting study had a large sample size and rigorous methodology, representing a comprehensive account on the epidemiology of dioxin.

The significance of the results of the study of Fingerhut and colleagues study was soon disputed by Collins, Acquavella and Friedlander (1992) on the basis that their conclusion about Soft Tissue Sarcomas (STS) was not reliable. Machamer and Douglas (1999) question the motivations of the authors critiquing the Fingerhut study. They state that all three worked for a chemical company responsible for dioxin pollution: "*One might expect that the company's interest in profit-making would determine which outcomes are acceptable in their employees' work.*" (p. 49). The social values of protection of the employer and the profit do not center in the debates. The conflicts of interest between the needs of the company and the needs of public health demand a closer look at the authors' reasoning and put into question the extent to which distinction can be clearly drawn between the epistemic (i.e. evaluation of reliability of data) and social aspects. In summary, "*the protracted debate on science and values has shown that it is deeply problematic to try and separate epistemic from non-epistemic, or cognitive from non-cognitive values*" (Wilholt, 2009, p. 96).

## 3.3 Generating a Framework on Scientific Aims and Values for Science Education

As the preceding coverage of aims and values in science illustrates, there is a range of values that philosophers of science have highlighted which are often debated at length (e.g. objectivity), making it difficult to provide a definitive summary. Research in science education can equally be extensive in its coverage of values such as those involved in the teaching and learning of scientific argumentation (e.g. Kolsto & Ratcliffe, 2008). Hence, it is particularly difficult to provide a normative set of aims and values that would be exhaustive and representative. Nevertheless, it is possible to develop a framework that can then be adapted for teaching and learning with the purpose of extending this framework to serve different educational goals. We recognize that the recommendations may shift or yield to other considerations in the context of the classroom. For instance, particular epistemic and cognitive values might be relevant to promote at different grade levels or at a particular cognitive ability level, while others might be visited repeatedly across all levels of schooling because they are more suitable for learners to acquire across age levels. The exploration of such implicit aspects of a framework on aims and values is subject to empirical investigation. The goal in this book is thus to be intentionally broad to provide a framework which can then be projected and extended. Two key questions can be posed that can be utilized by science education researchers as a toolkit in eliciting the types, the functions and the properties of values in science.

### 3.3.1 What Are Aims and Values in Science?

Through this question, we differentiate the epistemic, cognitive, social, political, cultural aims and values of science. The discussion in this chapter has highlighted primarily the range of epistemic and cognitive values that are inherent in science. Among these mentioned values - and admittedly, the review is not exhaustive - the following would be worthwhile to promote and include in science teaching and learning: consistency, simplicity, objectivity, empirical adequacy and novelty. Social, cultural and political contexts of science and their respective values are raised in Chap. 7 which include social norms such as being free from inductive bias, honesty, applicability to human needs and decentralization of power with respect to race and gender. A variation of the category of epistemic and cognitive values includes more nuanced and revised versions of traditional epistemic values. For example, the discussions about the relationship between objectivity and intersubjectivity would fit into this category. Achieving objectivity through intersubjective considerations by diverse community membership extends conventional notions of objectivity and subjectivity.

## 3.3.2 How Do Aims and Values Function in Science?

Through this question, the role that values can play in science is highlighted. For example, values can influence theory choice. They impact how scientists interact with their environments and affect methodological decisions and interpretations. Epistemic values such as empirical adequacy, accuracy, and explanatory power (Longino, 1995) function in conjunction with other values (e.g. cultural) to favor one theory over the other, or one set of findings over another. In this sense, they play a powerful role in the process of knowledge growth and development. The function and use of values in science can be complex. For instance, as we have seen through Irzik and Nola's (2011) discussion they can have interchangeable functions. These values become also aims in the sense that students would aim towards collecting accurate data and constructing more powerful explanations. These values are not about the "right answer" as much as about internalizing a set of understandings about how to conduct scientific inquiry or how to understand scientific inquiry from a holistic perspective. In the science classroom, emphasizing epistemic values signals to students the importance of conducting accurate measurement, recording accurate observations, and seeking 'higher' explanatory power. Creating and developing classroom cultures where the teacher and the students are explicitly aware of shared epistemic values will enable them to have a common language in approaching, conducting and interpreting scientific activities (Fig. 3.1).

Overall, the preceding questions and the review of literature lead to the conceptualization of aims and values from a range of epistemic and cognitive aspects. The range of values are represented as a corner of a triangle in order to signify that they are not always easily distinguishable but rather that the boundaries between them can be blurry and continuous. Because of emphasis in this chapter on epistemic and cognitive aims and values, for the sake of simplicity in communication, the social, political and cultural values are unpacked in Chap. 7. Furthermore, the representation intentionally excludes ethical and moral values. This is in agreement with Irzik and Nola's (2011) view that aims of science do not concern morality or ethics.

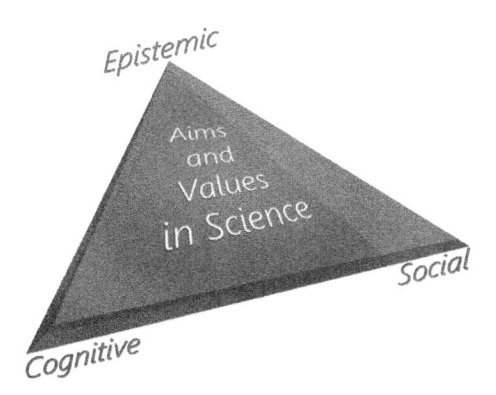

**Fig. 3.1** Aims and values in science

In other words, science is not an enterprise that aims to establish moral codes or address issues of ethics. In this sense, it is not an inherent ambition of science to deal with morality and ethics. However, scientists are expected to uphold particular ethical principles and moral codes such as honesty, and we do recognize that morality and ethics are important and can be intricately linked to the socio-political and cultural contexts of science. The relationship between moral and ethical values and social norms are discussed in Chap. 7.

## 3.4 Educational Applications

Science educators should pay attention to aims and values of science as discussed by a representative and select few philosophers of science such as Irzik and Nola (2011), Allchin (1999), Carrier (2013), and Longino (1990). We believe that this is important for at least the following reasons: (a) Inform students that scientific knowledge and scientific practices operate within an agreed upon set of values that guide scientific activity, (b) Support the acquisition of epistemic values through using instructional methods that facilitate deeper engagement with scientific content and practices, and (c) Raise awareness that aims and values can result in bias, for instance in terms of the design of an investigation or communication of scientific knowledge.

Based on Fig. 3.1, a framework can be derived that could be of use to science educators in summarizing, conceptualizing and visualizing key categories of the aims and values of science in a language that has pedagogical merit. For example, the epistemic goals can be considered to be related to goals that have to do with knowledge construction, evaluation and revision practices in the classroom. Hence, the epistemic dimension of the aims and values can be labeled as a 'knowledge' category in relation to the science curriculum. This category has several values associated with it, for instance objectivity, novelty and accuracy. (It is worthwhile to note that some philosophers of science including Thomas Kuhn, would prefer "predictive power" over 'accuracy'. For educational purposes, however, 'accuracy' seems more likely to be pedagogically and cognitively useful since the concept is already rather prevalent in school science.) The cognitive aims and values could be considered as aspects of 'reasoning' and can be promoted as those ways of thinking that highlight the key values in scientific reasoning. Some examples from earlier discussion are related themes such as revising convictions and critical examination. However from a philosophical perspective, it is at times difficult to distinguish between epistemic and cognitive values, in the sense that some aspects might be overlapping. For example, the value of objectivity can be both an epistemic aim as well as a cognitive pattern. As science educators, the purpose is neither to resolve such debates nor to contribute to them in substantial ways. Rather, the purpose is to use them in ways that can inform science education.

As Irzik and Nola (2014), the categories of epistemic and cognitive aims and values are collapsed to represent them together, although each particular value can be interpreted either as instances of both or either. The third category related to the social, political and cultural dimensions of the aims and values can be briefly

## 3.4 Educational Applications

referred to as the 'social' aspects and include examples such as addressing human needs, decentralizing power and honesty. As a guideline, a set of aims and values can be selected to be targeted in each category from which an educational application can be derived such as those illustrated in Table 3.1.

In order to illustrate the relevance of scientific aims and values and address the mentioned rationales in science education, three questions can be raised that help generate some concrete examples: (a) What are the cognitive and epistemic aims and values of science? (b) How do aims and values function in science? and (c) What are the properties of aims and values in science? These three questions are sufficiently broad to ensure that appropriate learning goals can be specified with respect to aims and values of science in science education. For each question, an example based on the theoretical discussions covered earlier in this chapter is provided. In each case, the transformation of the theoretical ideas for educational purposes can be illustrated, and their implementation at the practical level of the science lessons can be considered.

The example illustrated in Table 3.2 is related to question (a) in the previous paragraph and illustrates how a specified set of cognitive and epistemic aims and values of science can be applied in the context of a writing frame. Drawing from Helen Longino's (1990) work on procedural standards a set of values are selected such as "critical examination" and "addressing anomalies and counter instances". For each procedural standard, a rationale is developed for the epistemic value in the science lesson and example statements are provided that could be used to support students' writing. In other words, each procedural standard is indicated in a stem of a statement that can provide the support for students to appropriate it.

Another example targets the issue of how aims and values can play alternative functions in science. Irzik and Nola's (2011) depiction of how the value of simplicity can function as a rule, for instance in choosing between alternative theories, and also as an evaluative criterion in judging the explanatory power of a theory. A relevant instance to illustrate these concepts at the level of the classroom involves evidence that can be used to highlight the claim that day and night are caused by a spinning earth. One piece of evidence consists of a long exposure photograph of stars which appear to be going around concentric circles. At least two alternative theories can be proposed here to account for the circular trail of stars. One explanation could be that all the stars are rotating in a similar orbit. Another explanation could be that the earth revolves around its axis. Between these two alternative explanations, the value of simplicity would dictate choosing the second explanation over the first one. The value of simplicity can also illustrate why an explanation that relies on a lot of complex factors and assumptions, such as all stars behaving similarly, should be avoided. The idea that the circles are caused by a spinning earth provides a simpler account that has a better explanatory power than the alternative, which in fact invites more questions and raises a whole set of new assumptions about star behavior. In summary, the value of simplicity functions as a key criterion in judging the alternative explanations and informs decision-making about which explanation is likely to be taken seriously. Of course this piece of evidence will not be the sole evidence where such decisions about explanation choice are made. Other evidence could also be called into play, for instance the evidence from Foucault's

**Table 3.1** Application of epistemic-cognitive and social aims and values of science in science education

| | Aim/value | Educational application |
|---|---|---|
| Epistemic-cognitive | Objectivity | Seeking neutrality and avoiding bias |
| | Novelty | Searching for new explanations |
| | Accuracy | Ensuring that explanations are accurate |
| | Empirical adequacy | Basing claims on sufficient, relevant and plausible data |
| | Critical examination | Giving reasons to justify claims |
| | Addressing anomalies and counter instances | Recognizing opposite ideas and responding to objections |
| | Taking challenges seriously | Taking opposition to own ideas seriously |
| Social | Addressing human needs | Considering and respecting human needs |
| | Decentralizing power | Making sure nobody controls ideas to favor particular group biases |
| | Honesty | Being honest and acting honestly in all aspects of scientific activities |
| | Equality of intellectual authority | Respecting all ideas as long as they are evidence-based regardless of whose ideas they are |

3.4 Educational Applications

**Table 3.2** Example epistemic-cognitive aims and values in science based on Longino's (1990) procedural standards and suggested writing scaffold

| Procedural standard | Epistemic-cognitive value | Scaffold for students' writing |
|---|---|---|
| Critical examination | The student gives justified reasons that substantiate his/her claim | *I must give reasons for my claim because…* |
| Taking up criticism and responding to objections | The student recognizes the existence of counter ideas and provides response to objections | *I can see that my friend is against my idea. The way I would respond to her is….* |
| Addressing anomalies and counter instances | The student explains the anomaly between what they expected to find and what they actually found. If what they found what they expected, they would explain why the results did not fit the other group's expectations | *What I saw in the experiment was not what I expected. I would explain this by…* |
| Taking challenges seriously and trying to cope with them | The student takes challenges to difference in ideas seriously and tries to deal with them | *I should pay attention to my friend's idea and answer her by…* |
| Revisability of convictions | The student further develops his/her ideas and justify why they changed or elaborated them | *I think it's a good idea to change my idea because…* |
| Equality of intellectual authority | The student judges the merit of the claim by the extent to which it is supported by evidence | *I think group B's claim is more valid than our claim because…* |

Pendulum, in this example. This scenario can easily be translated into a lesson task through an argumentation framework of alternative claims (Erduran & Jimenez-Aleixandre, 2008). Students can be asked to propose alternative explanations to account for the pattern in the stars. Once they propose their explanations, the teacher could conduct a discussion to get the students to generate a set of criteria in order to evaluate which of these explanations are more likely to be true and why.

As part of discussions around the aims and values in science, the importance of understanding that not all values have the same properties should be stressed. Learning about values in science would necessitate that students understand the range of properties that values have. For example, some values are ultimate, and some are proximate (Allchin, 1999). As Longino (1990) identifies, values can be both constitutive and contextual. There are values that are intrinsic and extrinsic to science. Conventionally values were deemed to be extrinsic to science. While certain societal and cultural values may exist that do not relate to the scientific enterprise, science at large is influenced by and influences social and cultural values. In this respect, separation of the intrinsic and extrinsic values of science is not straightforward.

Such meta-level taxonomical understanding of how values are organized is likely to foster students' meaningful exposure to the complex issue of values in science. There are also aspects of values that might help expand knowledge or hinder knowledge production. Epistemic values are intrinsic to the conduct of scientific inquiry and to the production of valid scientific knowledge. It is inconceivable that scientific knowledge could be worthy of the name if it violated the values of accuracy or empirical adequacy. For example, the values of simplicity and explanatory power play important roles in theory-choice when two competing theories have an equally strong empirical backing, and in that sense these values determine what counts as valid. Hence values are necessarily intrinsic to science and can thus help develop or alternatively hinder science.

## 3.5 Fostering Scientific Aims and Values in Science Education

Example philosophical perspectives on aims and values in science have illustrated the potential for applications in science education. Examples were presented through which some of the scientific aims and values can be instilled in learners in classroom activities. Yet the establishment of aims and values of science in student discourse is a long-term commitment and raises questions about not only the adoption of these values but also their sustainability. On the one hand, a goal for science education is to ensure that a deep commitment to scientific aims and values are fostered in students and that students' identities are nurtured to internalize these aims and values. Some potential difficulties in the adoption of particular values are that the students might need to manage unconstructive values that are at odds with the scientific ones. For example, some students might be resistant to changing their own beliefs in light of new evidence hence not abiding with the value of revisability of convictions. Some others might not have developed yet the emotional competence to be able to deal with the consequences of abandoning their firmly held beliefs.

A key issue in the teaching of the aims and values in science concerns assessment. A traditionally unfamiliar set of goals in science education will necessitate that teachers themselves understand how to evaluate students' performance and attitudes about particular values in the classroom. Given the diffuse nature of values, it might be that teachers do not readily engage in taking values seriously as worthy of educational outcomes. Until aims and values in science become assessment goals, they are likely to be sidelined in classroom instruction. To facilitate incorporating aims and values, example sets of learning outcomes are provided that can be used by teachers as an indication of students' attainment of particular epistemic, cognitive and social values. Table 3.3 illustrates a formative assessment rubric that would facilitate not only the teaching and learning of aims and values in the science classroom but also provide a framework for their enactment. For example, in the case of the epistemic-cognitive value of objectivity, a high performing student can be expected to refer to the need to be cautious about bias and deliberately check data and claims for bias, whereas a 'satisfactory' indicator would be caution towards bias but not being consistent with being objective. A student who does not understand the need to seek objectivity and consistently demonstrates instances of bias would be considered in need of improvement relative to this goal.

3.5 Fostering Scientific Aims and Values in Science Education

**Table 3.3** Assessment rubric for epistemic, cognitive and social aims and values in science

| Aim/value | Target | Satisfactory | Needs improvement |
|---|---|---|---|
| *Seeking neutrality and avoiding bias* | Student is explicitly referring to the need to be cautious about bias and deliberately checks data and claims for bias | Student is somewhat aware of the need to be cautious about bias but is not consistent in ensuring objectivity | Student is not aware of the need to seek neutrality and demonstrates numerous instances of bias |
| *Searching for new explanations* | Student understands that science seeks new explanations because new explanations contribute to knowledge | Student understands that new explanations are desired by scientists but does not quite appreciate the significance of new knowledge in science | Student does not understand that scientists are not just interested in any knowledge. |
| *Ensuring that explanations are accurate* | Student strives to make sure that explanations that he/she provides are accurate for instance by evaluating explanations | Student recognizes that accurate explanations are important but does not engage in evaluation to ensure accuracy | Student does not understand that accuracy of his/her explanations are significant to consider |
| *Basing claims on sufficient, relevant and plausible data* | Student draws on data that are relevant and sufficient to justify claims | Student draws on data but does not engage in evaluating the relevance and plausibility of justification | Student does not understand that claims need to be justified by relevant, sufficient and plausible data |
| *Giving reasons to justify claims* | Student provides valid reasons for own claims | Student provides reasons for claims but the reasons are not valid | Student does not provide valid reasons for claims |
| *Recognizing opposite ideas and responding to objections* | Student understand an opposite point of view and can engage in a discussion in a constructive way | Student understands an opposite point of view but does not engage in a constructive discussion | Student does not understand an opposite point of view and does not engage with it |
| *Taking opposition to own ideas seriously* | Student recognizes the importance of taking opposition seriously and responds to it constructively | Student recognizes opposition but does not respond in a constructive manner | Student does not engage in opposition to own ideas |
| *Changing own ideas in light of evidence* | Student realizes the need to change own ideas in light of evidence and proceeds to do so | Student realizes that there is new evidence that disconfirms own ideas but does not engage with it | Student is not responsive to new evidence that challenge his/her ideas to change |
| *Considering and respecting human needs* | Student is able to appreciate that science needs to be respectful of human needs and acts with respect | Student understands that science has to respect human needs but does not engage in respect himself/herself | Student does not recognize that science has to address human needs and respect them |

(continued)

**Table 3.3** (continued)

| Aim/value | Target | Satisfactory | Needs improvement |
|---|---|---|---|
| *Making sure nobody controls ideas to favor particular group biases* | Student understands that the dominance of particular groups of people in the scientific enterprise might bias scientific practices and knowledge | Student understands that particular groups of people might dominate science but does not recognise the significance in terms of bias | Student does not understand that science might be controlled by particular groups of people |
| *Being honest and acting honestly in all aspects of scientific activities* | Student recognizes the importance of honesty in all aspects of science and strives to be honest | Student recognizes the importance of honesty in science but does not associate with it | Student does not understand the significance of honesty to science |
| *Respecting ideas as long as they are evidence-based irrespective of whose ideas they are* | Student respects others' ideas regardless of who they are | Student appreciates that everybody's ideas should be respected but does not practice it | Student respects only some people's ideas even though there might be good reasons to take others seriously |

A next step in the transformational work from theoretical ideas to educational frameworks would involve the interpretation of the levels in Table 3.3 for instructional purposes. In other words, each aim and value can be further elaborated to design actual activities that would elicit the expected value and thereby provide the context for its assessment. In future work, we intend to target the production, implementation and evaluation of the entire approach to envisaging science including the aims and values aspect.

It would be worthwhile to explore how the mentioned aims and values depicted in Table 3.3 could be related to science curriculum. To this effect, reference is made to the *Next Generation Science Standards* (NGSS) (NGSS Lead States, 2013) recently published in the USA to lead the next science education reform. This document includes an Appendix exclusively dedicated to "Nature of Science" (NOS) and highlights its key aspects in relation to the Practices and the Cross-Cutting Concepts parts of the document. The choice of this document is justified in its timeliness and also by the fact that the international science education community has been conventionally influenced by key policy documents published in the United States. For instance, the National Science Standards published in 1996 has been taken up by countless science educators from around the world in justifying the policy context of their research ranging from scientific inquiry to socio-scientific issues (Lee, Wu, & Tsai, 2009).

The NGSS coverage of NOS categories associated with Practices and Cross-Cutting Concepts is illustrated in Tables 3.2 and 3.3 respectively. At the middle school level, the category titled "Scientific investigations use a variety of methods" under scientific practices, states two understandings that explicitly target values: "*Science investigations are guided by a set of values to ensure accuracy of measurements, observations and objectivity of findings*" and "*Scientific values function as criteria in distinguishing between science and non-science*" (NGSS Lead States, 2013, p. 5). The precise nature of these values is not specified, although the reference to 'accuracy' and 'objectivity' is explicit and implies the consideration of the epistemic-cognitive values of science. The category titled "Science is a human endeavor", under cross-cutting concepts, includes the understanding that "*Men and women from different social, cultural and ethnic backgrounds work as scientists*" at the middle school level. Likewise in the same category, the understanding that "*Individuals and teams from many nations and cultures have contributed to science and advanced in engineering*", acknowledges the role of nationality and culture in science and engineering. The ways in which different nationalities, genders, ethnicities, cultures and political systems might have influenced science are not highlighted as a theme (Tables 3.4 and 3.5).

## 3.6 Conclusions

While the debate about the precise nature of the epistemic, cognitive and social aims and values continues among philosophers of science as well as other academics who take science as the object of their investigation (e.g. anthropologists, sociologists, linguists), there are particular assumptions that science and scientists make that are often not expressed explicitly in science lessons. There is the commitment to logic, rationality, skepticism and evidence. None of the illustrated aims and values are likely to find a place in the classroom until and unless some of these central tenets of science are also fostered and adapted. If science education fails to instill in learners those values, it is unlikely that students will emerge from schooling as scientifically literate citizens who possess scientific habits of mind. Without a deep sense of values to respect and accept that these commitments are significant enough to possess, science learners are bound to miss out on the fundamental aspects of science as a way of knowing.

Yet the aforementioned fundamental assumptions of science could also face resistance, which makes science teaching and learning seem like lost causes. In reflecting on his book titled *Moral Landscape* (Harris, 2012) in various multimedia presentations, Sam Harris has thoughtfully asked:

> If someone doesn't value evidence, what evidence are you going to provide that proves they should value evidence. If someone doesn't value logic, what logical argument would you invoke to prove he [sic] should value logic? (Sam Harris, personal communication with Erduran on May 6, 2013)

A potential difficulty apart from respect for evidence and logic is the criticism that the assumptions that science is based on are themselves psychological constructs and thus open to bias and interpretation, making science less than a reliable way of knowing. For instance, one could argue that skepticism and rationality are simply psychological states of particular groups of individuals who might be in the business of suppressing and controlling other people who subscribe to different psychological orientations. Likewise the value of rationality is not something that can be testable and thus evades the scrutiny that is so prized by science in the first place.

Ultimately it is the task of science education to ensure that learners can be supported in understanding the rationale for acquiring respect for evidence, logic, skepticism and rationality. An important aspect of this rationale is that it should be

**Table 3.4** Nature of science associated with practices in the *Next Generation Science Standards* (NGSS Lead States, 2013, p. 5)

Understandings about the nature of science

| Categories | K-2 | 3–5 | Middle school | High school |
|---|---|---|---|---|
| **Scientific investigations use a variety of methods** | Science investigations begin with a question | Science methods are determined by questions | Science investigations use a variety of methods and tools to make measurements and observations | Science investigations use diverse methods and do not always use the same set of procedures to obtain data |
| | Science uses different ways to study the world | Science investigations use a variety of methods, tools, and techniques | Science investigations are guided by a set of values to ensure accuracy of measurements, observations, and objectivity of findings | New technologies advance scientific knowledge |
| | | | Science depends on evaluating proposed explanations | Scientific inquiry is characterized by a common set of values that include: logical thinking, precision, open-mindedness, objectivity, skepticism, replicability of results, and honest and ethical reporting of findings |
| | | | Scientific values function as criteria in distinguishing between science and non-science | The discourse practices of science are organized around disciplinary domains that share exemplars for making decisions regarding the values, instruments, methods, models, and evidence to adopt and use |
| | | | | Scientific investigations use a variety of methods, tools, and techniques to revise and produce new knowledge |

3.6 Conclusions

| Scientific knowledge is based on empirical evidence | Scientists look for patterns and order when making observations about the world | Science findings are based on recognizing patterns | Science knowledge is based upon logical and conceptual connections between evidence and explanations | Science knowledge is based on empirical evidence |
|---|---|---|---|---|
| | | | | Science disciplines share common rules of evidence used to evaluate explanations about natural systems |
| | | Science uses tools and technologies to make accurate measurements and observations | Science disciplines share common rules of obtaining and evaluating empirical evidence | Science includes the process of coordinating patterns of evidence with current theory |
| | | | | Science arguments are strengthened by multiple lines of evidence supporting a single explanation |
| Scientific knowledge is open to revision in light of new evidence | Science knowledge can change when new information is found | Science explanations can change based on new evidence | Scientific explanations are subject to revision and improvement in light of new evidence | Scientific explanations can be probabilistic |
| | | | | Most scientific knowledge is quite durable but is, in principle, subject to change based on new evidence and/or reinterpretation of existing evidence |
| | | | The certainty and durability of science findings varies | |
| | | | Science findings are frequently revised and/or reinterpreted based on new evidence | Scientific argumentation is a mode of logical discourse used to clarify the strength of relationships between ideas and evidence that may result in revision of an explanation |

(continued)

**Table 3.4** (continued)

Understandings about the nature of science

| Categories | K-2 | 3–5 | Middle school | High school |
|---|---|---|---|---|
| **Science models, laws, mechanisms, and theories explain natural phenomena** | Science uses drawings, sketches, and models as a way to communicate ideas | Science theories are based on a body of evidence and many tests | Theories are explanations for observable phenomena | Theories and laws provide explanations in science, but theories do not with time become laws or facts |
|  | Science searches for cause and effect relationships to explain natural events | Science explanations describe the mechanisms for natural events | Science theories are based on a body of evidence developed over time | A scientific theory is a substantiated explanation of some aspect of the natural world, based on a body of facts that has been repeatedly confirmed through observation and experiment, and the science community validates each theory before it is accepted. If new evidence is discovered that the theory does not accommodate, the theory is generally modified in light of this new evidence |
|  |  |  | Laws are regularities or mathematical descriptions of natural phenomena | Models, mechanisms, and explanations collectively serve as tools in the development of a scientific theory |
|  |  |  | A hypothesis is used by scientists as an idea that may contribute important new knowledge for the evaluation of a scientific theory | Laws are statements or descriptions of the relationships among observable phenomena |
|  |  |  | The term "theory" as used in science is very different from the common use outside of science | Scientists often use hypotheses to develop and test theories and explanations |

3.6 Conclusions

**Table 3.5** Nature of science associated with cross-cutting concepts in the *Next Generation Science Standards* (NGSS Lead States, 2013, p. 6)

Understandings about the nature of science

| Categories | K-2 | 3–5 | Middle school | High school |
|---|---|---|---|---|
| **Science is a way of knowing** | Science knowledge helps us know about the world | Science is both a body of knowledge and processes that add new knowledge | Science is both a body of knowledge and the processes and practices used to add to that body of knowledge | Science is both a body of knowledge that represents a current understanding of natural systems and the processes used to refine, elaborate, revise, and extend this knowledge |
| | | Science is a way of knowing that is used by many people | Science knowledge is cumulative and many people, from many generations and nations, have contributed to science knowledge | Science is a unique way of knowing and there are other ways of knowing |
| | | | Science is a way of knowing used by many people, not just scientists | Science distinguishes itself from other ways of knowing through use of empirical standards, logical arguments, and skeptical review |
| | | | | Science knowledge has a history that includes the refinement of, and changes to, theories, ideas, and beliefs over time |

(continued)

**Table 3.5** (continued)

Understandings about the nature of science

| Categories | K-2 | 3-5 | Middle school | High school |
|---|---|---|---|---|
| **Scientific knowledge assumes an order and consistency in natural systems** | Science assumes natural events happen today as they happened in the past | Science assumes consistent patterns in natural systems | Science assumes that objects and events in natural systems occur in consistent patterns that are understandable through measurement and observation | Scientific knowledge is based on the assumption that natural laws operate today as they did in the past and they will continue to do so in the future. |
| | Many events are repeated | Basic laws of nature are the same everywhere in the universe | Science carefully considers and evaluates anomalies in data and evidence | Science assumes the universe is a vast single system in which basic laws are consistent |
| **Science is a human endeavor** | People have practiced science for a long time | Men and women from all cultures and backgrounds choose careers as scientists and engineers | Men and women from different social, cultural, and ethnic backgrounds work as scientists and engineers | Scientific knowledge is a result of human endeavor, imagination, and creativity |
| | Men and women of diverse backgrounds are scientists and engineers | Most scientists and engineers work in teams | Scientists and engineers rely on human qualities such as persistence, precision, reasoning, logic, imagination and creativity | Individuals and teams from many nations and cultures have contributed to science and to advances in engineering |
| | | Science affects everyday life | Scientists and engineers are guided by habits of mind such as intellectual honesty, tolerance of ambiguity, skepticism and openness to new ideas | Scientists' backgrounds, theoretical commitments, and fields of endeavor influence the nature of their findings |
| | | Creativity and imagination are important to science | Advances in technology influence the progress of science and science has influenced advances in technology | Technological advances have influenced the progress of science and science has influenced advances in technology |
| | | | | Science and engineering are influenced by society and society is influenced by science and engineering |

## 3.6 Conclusions

| Science addresses questions about the natural and material world | Scientists study the natural and material world | Science findings are limited to what can be answered with empirical evidence | Scientific knowledge is constrained by human capacity, technology, and materials | Not all questions can be answered by science |
|---|---|---|---|---|
| | | | Science limits its explanations to systems that lend themselves to observation and empirical evidence | Science and technology may raise ethical issues for which science, by itself, does not provide answers and solutions |
| | | | Science knowledge can describe consequences of actions but is not responsible for society's decisions | Science knowledge indicates what can happen in natural systems—not what should happen. The latter involves ethics, values, and human decisions about the use of knowledge |
| | | | | Many decisions are **not** made using science alone, but rely on social and cultural contexts to resolve issues |

presented from their perspective such that they can appreciate why they have to develop respect for evidence, logic and rationality. The adverse effects of the lack of such values could be one way of pointing out to students their merit. For example, case studies on forensic science and detective work could present accessible scenarios where individuals could be wrongly accused of crimes they did not commit if objectivity, evidence, skepticism and rationality are not employed. Students' disposition and interest in science are likely to improve once they are able to recognize the significance of upholding the various aims and values of science.

## References

AAAS. (1989). *Science for all Americans*. Washington, DC: American Association for the Advancement of Science.
Allchin, D. (1999). Values in science. *Science & Education, 8*, 1–12.
Carrier, M. (2013). Values and objectivity in science: Value-ladenness, pluralism and the epistemic attitude. *Science & Education, 22*, 2547–2568.
Collins, J. J., Acquavella, J. F., & Friedlander, B. (1992). Reconciling old and new findings on dioxin. *Epidemiology, 3*(1), 65–69.
Douglas, H. (2000). Inductive risk and values in science. *Philosophy of Science, 67*(4), 559–579.
Erduran, S., & Jimenez-Aleixandre, M. P. (Eds.). (2008). *Argumentation in science education: Perspectives from classroom-based research*. Dordrecht, The Netherlands: Springer.
Fingerhut, M. A., Halperin, W. E., Marlow, D. A., Piacitelli, L. A., Honchar, P. A., Sweeney, M. H., et al. (1991). Cancer mortality in workers exposed to 2,3,7,8-Tetrachlorodibenzo-P-Dioxin. *New England Journal of Medicine, 324*, 212–218. doi:10.1056/NEJM199101243240402.
Gluud, L. L. (2006). Bias in clinical intervention research. *American Journal of Epidemiology, 163*(6), 493–501.
Harding, S. (1991). *Whose science? Whose knowledge? Thinking from women's lives*. Ithaca, NY: Cornell University Press.
Harris, S. (2012). *The moral landscape*. London: Bantam Press.
Hempel, C. G. (1965). *Aspects of scientific explanation and other essays in the philosophy of science*. New York: Free Press.
Irzik, G., & Nola, R. (2011). A family resemblance approach to the nature of science. *Science & Education, 20*, 591–607.
Irzik, G., & Nola, R. (2014). New directions for nature of science research. In M. Matthews (Ed.), *International handbook of research in history, philosophy and science teaching* (pp. 999–1021). Dordrecht, The Netherlands: Springer.
Kolsto, S. D., & Ratcliffe, M. (2008). Social aspects of argumentation. In S. Erduran & M. P. Jimenez-Aleixandre (Eds.), *Argumentation in science education: Perspectives from classroom-based research* (pp. 117–136). Dordrecht, The Netherlands: Springer.
Kuhn, T. S. (1977). *The essential tension*. Chicago: University of Chicago Press.
Lee, M.-H., Wu, Y.-T., & Tsai, C.-C. (2009). Research trends in science education from 2003 to 2007: A content analysis of publications in selected journals. *International Journal of Science Education, 31*(15), 1999–2020.
Longino, H. (1990). *Science as social knowledge: Values and objectivity in scientific inquiry*. Princeton, NJ: Princeton University Press.
Longino, H. E. (1993). Essential tensions—Phase two: Feminist, philosophical, and social studies of science. In L. M. Antony & C. Witt (Eds.), *A mind of one's own. Feminist essays on reason and objectivity* (pp. 257–272). Boulder, CO: Westview Press.
Longino, H. (1995). Gender, politics and the theoretical virtues. *Synthese, 104*, 383–397.

# References

Longino, H. (1997). Feminist epistemology as a local epistemology. *Aristotelian Society Supplementary Volume, 71*(1), 19–36.
Longino, H. E. (2002). *The fate of knowledge*. Princeton, NJ: Princeton University Press.
Machamer, P., & Douglas, H. (1999). Cognitive and social values. *Science & Education, 8*, 45–54.
NGSS Lead States. (2013). *Next generation science standards: For states by states*. Appendix H. Retrieved from http://www.nextgenscience.org/next-generation-science-standards
Popper, K. R. (1957). Science: Conjectures and refutations. In *Conjectures and refutations. The growth of scientific knowledge* (pp. 43–78). London: Routledge, 1963, 2002.
Popper, K. R. (1963). *Conjectures and refutations*. London: Routledge and Kegan Paul.
Popper, K. R. (1975). *Objective knowledge*. Oxford, UK: Clarendon.
Resnick, D. (2007). *The price of truth: How many affects the norms of science*. Oxford, UK: Oxford University Press.
van Frassen, B. (1980). *The scientific image*. Oxford, UK: Clarendon.
von Glasersfeld, E. (1989). Cognition, construction of knowledge and teaching. *Synthese, 80*, 121–40.
Wilholt, T. (2009). Bias and values in scientific research. *Studies in History and Philosophy of Science, 40*, 92–101.

# Chapter 4
# Scientific Practices

The purpose of this chapter is to explore scientific practices and discuss their implications for science teaching and learning. In the past few decades, there has been increasing interest in the notion of science-as-practice. Curriculum reform documents such as the *Next Generation Science Standards* in the USA are increasingly advocating the teaching and learning of scientific practices. What are scientific practices? Why are scientific practices important to consider in science education? What heuristics can be developed to facilitate the teaching and learning of scientific practices in science lessons? These questions will guide the discussions in this chapter. The chapter focuses on three examples of scientific practices that are prevalent in international science curricula: classification, observation, and experimentation. The discussion gives rise to a heuristic that captures the relationships among the cognitive, epistemic and discursive practices of science. Implications of the heuristic to science education policy and instruction are presented.

## 4.1 Introduction

School science has been dominated by what seems to be an 'essential tension' between two competing curriculum emphases: one focusing on the products of science in the form of propositional knowledge of particular theories, laws and models, and another focusing on scientific processes. Problems associated with the first type of emphasis is rooted in the teaching of products of science in a disconnected fashion without giving learners a sense of the relations between different forms of scientific knowledge; how scientific knowledge grows; and what criteria, standards and heuristics drive growth of scientific knowledge (See Chap. 6 for further discussion on the nature of scientific knowledge in school science). Problems associated with the second type of emphasis, have led to teaching isolated science process skills, losing track of how they relate to one another and how they function within a larger set of scientific practices to yield meaningful scientific knowledge. The outcomes of oscillating between such emphases are: (a) reinforcement of an artificial separation between scientific products and processes, and (b) oversimplification of the nature of scientific knowledge and practices. As Schwab (1962) pointed out many decades ago, students need to understand both the substantive and the syntactic structures of science. The substantive structure refers to "a body of concept-commitments about the nature of the subject matter functioning as a guide to inquiry", while the syntactic structure refers to "the pattern of the discipline's procedure, its method, how it goes about using its conceptions to attain its goal" (Schwab, p. 203). Communicating both structures in curriculum and instruction is a desirable goal in science education.

The primacy of "science-as-knowledge" has been challenged since the 1970s with increasing attention devoted to "science-as-practice" (e.g. Pickering, 1992; Rouse, 2002). Underlying the recent debate is the notion that science cannot be viewed merely as a body of knowledge but rather as a particular epistemic, social and cultural practice. The work of David Bloor, Stephen Shapin, Karin Knorr Cetina, Bruno Latour and Steve Woolgar, among others have contributed to rich scholarship that capitalized on the philosophy and sociology of scientific practices. Contrary to the predominant emphasis on scientific knowledge, the articulation of scientific practices aimed to emphasize those social and cultural processes (such as peer review and norms of research groups) that constitute and underpin the generation, evaluation and revision of scientific knowledge. The significance of this line of work for science education is that it highlights the necessity of teaching science in a more holistic context which justifies from the learners' point of view the processes as well as the products of the scientific endeavor.

While investigating questions in the natural world, scientists gather, organize and classify data in order to formulate knowledge. In science, knowledge creation has traditionally followed the path of systematic exploration, observation, description, experimentation and analysis all conducted within the communication framework of a specialized research community with its accepted methodology (Kwasnik, 1999). The process, however, is not entirely rational. Often hunches, values and insight are involved in the processes of scientific inquiry (Bronowsky, 1978). There are also particular political, cultural (e.g. Olson, 1998) and economic (e.g. Irzik, 2010) factors that influence scientific inquiry. These underlying political, cultural, economic, social factors that might come into play in scientific inquiry and eventual formulation of scientific knowledge are often sidelined as learning outcomes in school science.

In terms of processes of scientific inquiry, science curricula place considerable emphasis on classification, observation and experimentation. Yet, these activities tend to be rather limited in terms of their epistemic framing. In other words, while they are addressed to varying depths in science lessons, there is limited questioning and discussion as to how they contribute to scientific knowledge and its growth. Their coverage tends to be disparate without attention to how they operate within a coherent set of practices that integrate epistemic, cognitive and social-institutional dimensions of science. Often they are procedural in their implementation in science lessons where students are instructed to conduct investigations that are mostly closed-ended and formulaic in nature (Chinn & Malhotra, 2002). Not only do these inquiries tend to be of the simple kind, they are not typically related to the epistemic development of scientific knowledge. Likewise, the infusion of the social, cultural and institutional factors that co-constitute scientific knowledge and its development are often overlooked.

The chapter targets three activities (i.e. classification, observation and experimentation) to illustrate how they can be conceptualized as examples of "scientific practices" contributing to the generation of scientific knowledge. Through the description and positioning of these activities as "scientific practices" an argument is developed for the infusion of the epistemic, cognitive and social-institutional aspects of the scientific enterprize. The discussion leads to the generation of a heuristic that can provide conceptual and pedagogical coherence in how scientific practices can be taught and learned to address the limitations of the conventional instructional coverage of scientific processes. The chapter concludes by relating the proposed heuristic to the notion of "scientific practices" within the USA's *A Framework for K-12 Science Education* (NRC, 2012) in order to illustrate its relevance to policy contexts as well as its potential use as a pedagogical tool.

## 4.2 Differentiating Scientific Practices, Processes and Activities

At this point, three related concepts – processes, activities and practices – need to be differentiated. These concepts are sometimes used to denote similar ideas but are situated in fundamentally different theoretical assumptions. Although 'processes', 'activities' and 'practices' have often been used interchangeably to refer to aspects of science such as experimentation and observation, their precise attributes are guided by their broader theoretical assumptions about science as well as science learning. The term "scientific processes" typically refers to how scientific research is done. Earlier attempts to transfer this important idea of engaging students with scientific processes resulted in simplifying those processes to smaller chunks: science process skills. The notion of "science process skills" was very much influenced by the positivist characterizations of science (e.g. Dillashaw & Okey, 1980) and, in its limited focus, tended to emphasize particular skills such as manipulation of variables and interpretation of graphs. On the other hand, "scientific practices" intend to situate these aspects of science into broader epistemic and discursive practices such as "making sense in patterns of data" and "coordination of theory and evidence" (e.g. Sandoval & Millwood, 2005).

In their original paper, Irzik and Nola's (2011) refer to "scientific activities" in the following way:

> Observing and experimenting are clearly scientific activities, hence the category 'activities'. As we have pointed out before, when described at a very general level, observing will be an activity common to all sciences. However, it should be noted that observational practices will obviously vary according to the scientific discipline in which observation is carried out. For example, specific observational skills are required in observing the planets and stars using telescopes; they involve being able to position the heavenly bodies against the cross-wires of a telescope while simultaneously noting the time on a clock to avoid the biases associated with what is known as an observer's 'personal equation'. These are quite different from the recognitional skills of, say, fossil prospectors in Northern Kenya and Ethiopia who become very good indeed in identifying fossils on the ground from other rocks. (p. 597)

The authors further differentiate scientific practices from "material practices" which are instances such as "calibrating scientific instruments and planning, setting up and carrying out experiments" (Irzik & Nola, 2011, p. 597). Scientists may also resort to "mathematical practices" which could range from "methods for applying equations of dynamics to some concrete case of motion such as a swinging pendulum" (Irzik & Nola, 2011, p. 597). Given the authors' disciplinary affiliation in philosophy of science, the reference to processes is not similar to the science education researchers' depictions as described earlier. Irzik and Nola's approach highlights the disciplinary variations as well as similarities between scientific 'practices':

> There are a set of activities that are characteristics for some sciences but not others, thereby forming a family resemblance set. These include observational practices, material practices, mathematical practices and so on. Looked at broadly all sciences will have observational practices (more broadly, data collection practices), but from a more fined grained point of view, the observational practices involved in, say, astronomy will not be the same as in ethology or archeology. Again, physics will involve both material and mathematical practices extensively, while for botany there are classificatory practices but there is little, if at all, mathematical practices. And so on for all the individual sciences; each will draw on some sub-set of characteristics but not others. (Irzik & Nola, 2011, p. 597)

In their subsequent iteration of the family resemblance approach within the *Handbook of History, Philosophy, Sociology of Science and Science Teaching* edited by Michael Matthews, Irzik and Nola (2014) replace reference to the word 'activities' with "processes of inquiry". Their conception of "processes of scientific inquiry" constitutes a significantly different theoretical position from those of "science process skills" often portrayed in science education.

In this book, the term 'practices' is used because of (a) its presence within the science education research literature primarily in reference to epistemic practices (e.g. Jimenez-Aleixandre & Reigosa, 2006; Krajcik & Merritt, 2012; Sandoval & Millwood, 2005), and (b) its contemporary currency within the science education curricular policy, for instance in terms of its prominence within the *Next Generation Science Standards* in the USA. The theoretical articulation of 'practices' incorporates the epistemic variations that might exist between different branches of science which can be important to promote in science learning (e.g. Erduran, 2007). Construing scientific activities as practices is not a mere term substitution or preference, but involves substantial reconceptualization of how scientific activities become epistemic entities, contributing to the generation and evaluation of scientific knowledge.

An example will illustrate how we envisage "scientific practices" as being different from activities and processes. In school science, classification is mostly addressed as a sorting activity or a tool for organizing observations with little or no attention given to its explanatory/predictive power or to how it fits within a broader theoretical framework. For instance, students might be asked to classify objects for which there is no broader theoretical significance, such as sorting out buttons and pencils. This sense of classification could be considered as an activity. This is in sharp contrast to how scientists use classification not only to organize existing relationships but also predict new ones all the while operating within a broader theoretical framework. Another example from chemistry is how Mendeleev's classification of elements on the basis of periodicity led to the prediction of Gallium hence highlighting the role that classification can play in predictions. Conceiving of classification as practice in science education, lifts the level of engagement with it from being an isolated activity to one that is situated in the broader epistemic, cognitive, and social-institutional practices of the discipline.

Scientific practices involve not only the epistemic but also the social-institutional and cultural components that underlie choices made within the enactment of activities. For example, scientists engage in experimentation whereby particular results are derived through controlled trials that are negotiated and discussed within teams of researchers relative to particular evaluative criteria, and reviewed by peers for wider communication. Scientific practices further include the conceptual and theoretical elements that underlie the choice of tools that are deployed in their constitution. They underscore the discursive relationship between the practices themselves and the individuals and communities by whom they are being practised. Situating activities or processes such as classification and experimentation within the broader practices transforms them from mere discrete and isolated activities or processes to grounded practices. The scientific practices involve the collection of data for particular purposes, for instance modeling of phenomena. They involve the coordination of evidence and models through discursive processes such as argumentation. The practices are thus interdependent on one another and service the generation of scientific knowledge. In summary, embeddedness in broader theoretical frameworks and interconnectedness in epistemic, cognitive and social-institutional mechanisms are the defining features of scientific practices.

## 4.3 Examples of Scientific Practices: Classification, Observation and Experimentation

There are numerous activities that underlie scientific practices. These include observation, classification, and experimentation which are discussed and elaborated in an effort to highlight some key features. To begin with, observation is a central scientific activity. Some scientists make direct observations of phenomena in the natural and physical world, such as the task of a botanist who studies plant species through direct observation or an astronomer who studies distant galaxies using special tools like telescopes and mathematical models. The number of twentieth century scientists, philosophers and cognitive scientists who have contemplated the nature of observation and the implications for science is enormous. Within philosophy of science, observation has centered quite strongly in discussions about the nature of truth, and indeed much debate has been generated through articulation of some fundamental questions such as the following: What is the role of observation in getting to know the physical world? What is the relationship between human perception and the real world? With the advent of developments in cognitive psychology during the twentieth century, the mind, the idea of consciousness and human behavior were critically examined from a variety of perspectives with implications for how philosophers' accounts of science correspond to cognitive psychological accounts.

For example, cognitive scientist Nancy Nersessian's work has focused on the use of original historical accounts to examine how model-based reasoning functions in science (e.g. Nersessian, 2003). While some cognitive scientists examined perception and language as elemental parts of conscious awareness, others felt they could not be extracted from the whole of consciousness and analyzed independently, for instance through language analysis. Yet others considered perception as an innate conversion of sense data into linguistic code. Bertrand Russell asserted that there are fundamental elements of language just as there are fundamental elements of nature (Russell, 1996/1912). Language was the link between perception and understanding. Russell also argued that there was an innate link between sensory experience and the physical world.

Cognitive scientists, like Philip Johnson-Laird, argue that perception, ideas and beliefs are all treatable as mental representations or symbols. Johnson-Laird describes the "phenomenological experience of the world" as

> ...a triumph of natural selection. We seem to perceive the world directly, not a representation of it. Yet this phenomenology is illusory: what we perceive depends on both what is in the world and what is in our heads—on what evolution has 'wired' into our nervous systems and what we know as a result of experience. (Johnson-Laird, 1993, p. 471)

Observations underscore the data that scientists use to generate models, theories and laws. Irzik and Nola (2011) distinguish between observational and experimental data. While they do not explicate in detail these notions in this particular paper, Gurol Irzik has informally provided the following distinction:

> Each observational and experimental data can be expressed in terms of a statement of the form: such and such object has such and such property. For that reason, statements that express observational and experimental data are singular statements, which are different from scientific laws (such as PV = constant) that are expressed in terms of universal statements. Given all this, both observational and experimental data, provided they contain no errors, typically function as evidence for or against theories or hypotheses, they are used in scientific explanations and often called initial conditions and thus constitute part of the

## 4.3 Examples of Scientific Practices: Classification, Observation and Experimentation

corpus of scientific knowledge. In his *Logic of Scientific Discovery*, Popper discusses these points at some length. See especially Chapter III. (Irzik, electronic communication with Erduran, January 24, 2013)

Irzik further explains that observational reports concern the data that are obtained through observation, for example, the data obtained with a telescope on a planet at different times and different locations. Experimental data on the other hand, are obtained through experiment. For example, the measurements that enabled Boyle to find out about the law that carries his name (as expressed by PV = constant). Broadly speaking, both forms of data could count as observations. Science education literature has captured some of the debates within philosophy of science (e.g. Hodson, 1998; Matthews, 1994; Norris, 1985) illustrating key themes such as the theory-ladenness of observations, and observation-explanation dichotomy. One strategy to include these debates in science education could involve the unpacking of the kinds and properties of observations as well as their links to theory, explanation and instrumentation. When embedded in scientific theories and interlinked to other epistemic practices such as modeling, observation becomes a scientific practice. This sense of observation distinguishes it from the generic and all-encompassing activity of humanity that relies on understanding the world through sensory experience.

A second example of a scientific practice is classification. Classification is utilized in many science disciplines, for example the classification of species in biology and the periodic arrangement of elements in chemistry. Classification "is the meaningful clustering of experience" (Kwasnik, 1999, p. 24) and it can be used in a formative way during the preliminary stages of inquiry as a heuristic tool in discovery, analysis, and theorizing (Davies, 1989). It can facilitate the process of knowledge generation. Classification operates through particular structures such as hierarchies and sets. Understanding of hierarchical classification dates back to Aristotle (Ackrill, 1963) who argued that all of nature was a unified whole which could be sub-divided into 'natural' classes, and each class further into sub-classes. Aristotle posited that only exhaustive observation can reveal each entity's true attributes and only philosophy can guide in the determination of the necessary and sufficient attributes for membership in any given class.

According to Kwasnik (1999, pp. 25–26) hierarchies have strict structural requirements that are summarized in the first column of Table 4.1. Kwasnik illustrates one example hierarchy in the context of medical science. Eye diseases are classified at a first level as conjunctival and corneal diseases. Conjunctival diseases in turn have subsets of conjunctival neoplasm and conjunctivitis. Conjunctivitis is then unpacked into allergic, bacterial, ophthalmia neonatorum and trachoma as well as viral, keratoconjunctivitis and Reiter's disease, and so on. The idea of a hierarchy in classification is illustrated by the biology example drawn from TutorVista.Com, a popular website that provides instructional resources and tutorials to secondary school students (Fig. 4.1). Classification tools using a hierarchical organizational structure similar to the ones suggested by Kwasnik are familiar to science educators through the work of Joseph Novak who developed them in the form of concept

Table 4.1 Requirements for hierarchy (From Kwasnik, 1999, pp. 25–26) and a biology example

| Requirements for hierarchy (verbatim from Kwasnik) | Biology example |
| --- | --- |
| **Inclusiveness**: The top class is the most inclusive class and describes the domain of the classification. Top class includes all its subclasses and sub-subclasses | Top class is "Animal Tissues" |
| **Species/differentia**: A true hierarchy has only one type of relationship between its super- and subclasses and this is the generic relationship also known as species/differentia, or more colloquially as the is-a relationship | The blood is a kind of fluid which is a kind of connective tissue which in turn, is a kind of animal tissue |
| **Inheritance**: This requirement of strict class inclusion ensures that everything that is true for entities in any given class is also true for entities in its subclasses and sub-subclasses. This property is called inheritance, that is, attributes are inherited by a subclass from its superclass | Whatever is true about connective tissues is also true of skeletal tissues, which in turn are true of cartilage |
| **Transitivity**: Since attributes are inherited, all sub-subclasses are members of not only their immediate superclass but of every superclass above that one. This property is called transitivity | If bone is a skeletal tissue, and skeletal tissue is a connective tissue, then bone is a connective tissue |
| **Systematic and predictable rules for association and distinction**: The rules for grouping entities in a class (i.e. creating species) are determined beforehand, as are the rules for creating distinct subclasses (differentia). Thus all entities in a given class are like each other in some predictable (and predetermined) way, and these entities differ from entities in sibling classes in some predictable (and predetermined) way. They are differentiated from each other along some predictable and systematic criterion of distinction | Bone and cartilage are alike in that they are both kinds of skeletal tissue. They are differentiated from each other along some predictable and systematic way |
| **Mutual exclusivity**: A given entity can belong to only one class. This property is called 'mutual exclusivity' | Blood belongs to the fluid class. It cannot belong to fluid and the adipose class at the same time |
| **Necessary and sufficient criteria**: In a pure hierarchical classification, membership in a given class is determined by rules of inclusion known as necessary and sufficient criteria. To belong to the class, an entity must have the prescribed (necessary) attributes; if it has the necessary attributes, this then constitutes sufficient warrant, and the entity must belong to the class | Skeletal tissue is the bony, ligamentous, fibrous, and cartilaginous tissue forming the skeleton and its attachments. Cartilage is an attachment on the skeleton |

maps in 1972 to understand changes in children's knowledge of science (Novak & Cañas, 2008). Since then popularity of concept mapping tools has grown for organizing disciplinary knowledge in science curriculum planning, scaffolding student learning; identifying student science conceptions prior to and post instruction; capturing shifts in conceptual understanding in student thinking; and tools for understanding and comparing expert and naïve knowledge schema.

In Kwasnik's (1999) scheme, the requirements for hierarchy include concepts such as 'inclusiveness', 'inheritance' and 'transitivity'. These requirements are

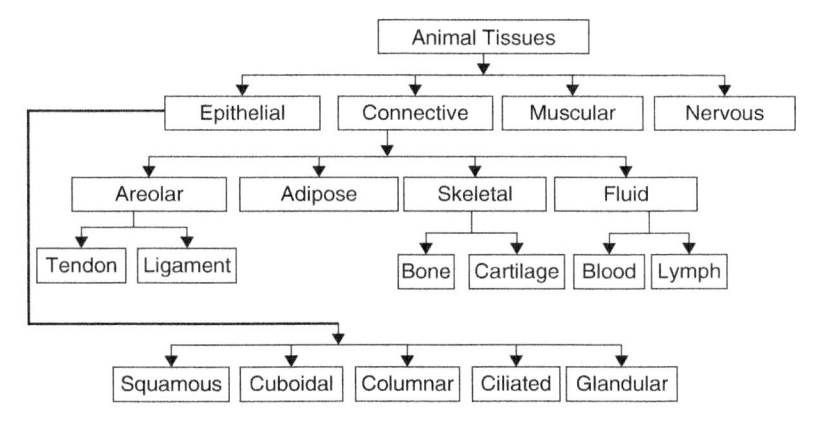

4.3 Examples of Scientific Practices: Classification, Observation and Experimentation 75

**Fig. 4.1** Classification of animal tissues (Reproduced from TutorVista.com)

essentially criteria that enable the evaluation of categories that can belong or not within the hierarchy. Often in school biology, for instance, hierarchies are introduced to students without any explicit reference to such criteria (i.e. inheritance, inclusiveness), which tend to be implicit. As an example instructional approach, each of these criteria can be phrased as questions that guide students in thinking about characteristics of a classification system, such as: Does my idea follow from the idea that is above in the set? (Inclusiveness) Does it belong to the idea that is in the higher set above it? (Transitivity)?

Kwasnik (1999) discusses other classification categories such as a tree and paradigms. However, concept maps and "tree structures" are not the only way to classify or represent knowledge. A good example from chemistry of a classification system is the Periodic Table of elements (Hjorland, Scerri, & Dupre, 2011). When the Periodic Table was first proposed, there was already a body of knowledge about individual elements such as atomic weight (Scerri, 2007). It was observed that elements could be arranged in a systematic order according to atomic weight, and this would show a periodic change of properties. This early Periodic Table proved to be a very useful tool, leading to the discovery of new elements and new understandings of already known elements. With the advent of the atomic theory, the Periodic Table was no longer just a descriptive classification system but also possessed predictive power for the yet undiscovered elements. Even though a new explanatory power has been attributed to the Periodic Table through the atomic theory, the original Periodic Table did not have to undergo fundamental changes in structure.

Scerri (2007) discusses Mendeleev's successful prediction of Gallium, which was unknown at his time. Citing the work of J. R. Smith (unpublished PhD thesis, University of London, 1975, pp. 357–359) he illustrates how Mendeleev's predicted and observed properties of Gallium compared. In Table 4.2, a few of these properties are selected to illustrate the incredible similarity in the actual observations. The Gallium example illustrates how the activity of classification was not only a

**Table 4.2** The predicted and observed properties of gallium

| Predicted | Observed |
|---|---|
| Properties should represent the mean of those of Zn and eka-silicon on the one hand, and those of Al and In on the other | Many properties do indeed represent a transition from those of Zn to those of Ge on the one hand, and from those of Al to those of In on the other |
| More acidic than eka-boron | More acidic than scandium |
| Atomic weight: ca. 68 (H=1) | Measured atomic weight: 69.2 (H=1) |
| The hydrous oxide will dissolve in KOH solution | The stable oxide is $Ga_2O_3$, gallic oxide. This is soluble in HCl, $H_2SO_4$ and aqueous alkalia hydroxide and ammonia but it was been previously strongly heated it dissolves in these media only extremely slowly |
| Specific gravity: ca. 6.0 (Atomic Volume: ca. 11.5) | Specific gravity: 5.9 (Atomic volume, 11.8) |
| Eka-aluminum is likely to be discovered spectroscopically (on the grounds of its expected volatility) like In and Tl | Gallium was indeed discovered spectroscopically |

From Scerri (2007, pp. 133–134)

descriptive account but also possessed predictive and explanatory power, providing explanations of chemical concepts such as acidity. The predictive power of classification schemes is one aspect of classification that is not sufficiently captured in science education, often situating classification as a mere descriptive organizational tool. The Gallium example can be used also as a basis for illustrating how Mendeleev successfully predicted not only this element but also Germanium and Scandium.

The preceding examples from biology and chemistry not only illustrate how classification can operate in science in broad terms but also they point to particular epistemic features of the classification as a *practice*. The reference to criteria for deciding whether a concept belongs to the hierarchy or not, or the role of the accumulated disciplinary knowledge in guiding the placement of an element in the Periodic Table indicate that classification in science is more than just sorting and describing ideas, objects, and relationships. Classification is a scientific practice constituted by an epistemic purpose. The neglect of the epistemic dimensions of classification in school science (e.g. the role of epistemic criteria in establishing classification systems; the explanatory and predictive power of classifications) reduces it to a sorting activity. This is because there is no reference to a broader theoretical context or purpose such as those illustrated in the animal tissue and Gallium examples. In the case of the animal tissue, the broader theoretical context is the cell theory while the gallium example relates to the periodicity concept.

The third scientific practice addressed in this section is experimentation. There is substantial amount of research on the history, philosophy and sociology of experimentation providing insight into how experimentation works in science (e.g. Latour & Woolgar, 1979; Mayo, 1996; Shapin & Schaffer, 1985). Kuhn (1977) claimed that the rise of modern physical science resulted from two simultaneous developments.

## 4.3 Examples of Scientific Practices: Classification, Observation and Experimentation

The first was the radical conceptual and world-view change that occurred in what he calls the classical, or mathematical sciences, such as astronomy, statics and optics. The second was the novel type of Baconian, or experimental, sciences that emerged, dealing with the study of light, heat, magnetism and electricity, among other things. Kuhn argued that it was not before the second half of the nineteenth century that a systematic interaction and merging of the experimental and mathematical traditions took place. An example is the transformation of the Baconian science of heat into an experimental-mathematical thermodynamics during the first half of the nineteenth century. At about the same time, the interactions between science and technology increased substantially.

Radder (2009) outlines two primary features of experimentation: intervention and reproducibility. In order to perform experiments, experimenters have to *intervene* actively in the material world; moreover, in doing so they *produce* all kinds of new objects, substances, phenomena and processes. Radder explains that experimentation involves the material realization of the experimental system as well as an active intervention in the environment of this system. Hence, a central issue for a philosophy of experiment is the question of the nature of experimental intervention and production, and their philosophical implications. Sometimes scientists devise and discuss so-called "thought experiments" (Brown, 1991). However, in such 'experiments' the crucial aspect of intervention and production is missing.

The notion of reproducibility is at the heart of debates on experimentation. Reproducibility states that a successful performance of an experiment by the original experimenter is an achievement that may depend on certain idiosyncratic aspects of a local situation (Radder, 2009). A purely local experiment that cannot be carried out in other experimental contexts will be unproductive for science. However, since the performance of an experiment is a complex process, no repetition will be strictly identical to the original experiment and many repetitions may be dissimilar in several respects. For this reason, *what* needs to be reproducible has to be specified. Furthermore, there is the question of *who* should be able to reproduce the experiment.

Another important topic in discussions on experimentation is the tendency to take the production of experimental knowledge for granted and to focus on theoretical knowledge (Radder, 2009). Yet science has been tightly linked to development of technology. Experiments make essential use of technological tools, and experimental research often contributes to technological innovations. Moreover, there are substantial conceptual similarities between experimental and technological processes, most significantly the implied possibility and necessity of the manipulation and control of nature (Radder, 2009). These kinds of issues have made science-technology relationship a central topic in the study of scientific experimentation.

According to Radder (2009), the relationship between experiment and theory is a significant aspect of scientific experimentation. He identifies a range of relationships. First, the production of theories as a result of experiments can be investigated (Franklin, 1986). Second, the role of existing theories, or theoretical knowledge, within experimental practices can be examined. At one extreme, it can be claimed that experimentation is theory-free. A more moderate view is that theory-free experiments are possible and do occur in scientific practice. This view admits that performing such 'exploratory' experiments does require some ideas about nature and apparatus, but not a well-developed theory about the phenomena.

Gooding, Pinch, and Schaffer (1993) focus on the instruments and equipment employed in experimental practice. Others (e.g. Radder, 2003) also have shown that the investigation of scientific instruments is a rich source of insights for a philosophy of scientific experimentation. For example, the role of visual images in experimental design could be investigated. According to Radder (2009), there are differences in the way that instruments can be characterized, for instance, instruments that represent a property by measuring its value (e.g. a device that registers blood pressure), instruments that create phenomena that do not exist in nature (e.g., a laser), and instruments that closely imitate natural processes in the laboratory (e.g. an Atwood machine, which mimics processes and properties of falling objects).

The preceding discussion of experimentation raises several themes for elaboration in science teaching and learning. Experimentation can be positioned as scientific practice rather than the conventional activity whereby students are instructed to follow prescribed procedures, dubbed as the 'cookbook' approach. Experimentation in science is not about predetermined set of procedures. Scientists often invent new procedures and approaches to conduct investigations to address research problems. The identification of the relevant and appropriate experimental procedures is as important a part of the discussions among scientists as the data, models, theories and laws (see Chap. 5). Positioning experimentation not as a procedural activity but rather as an important epistemic practice of science elevates its current mindless and procedural status in school science to scientific practice that relies on the use of epistemic criteria and standards. For instance, the case of reproducibility, the link between experiment and theory, intervention, and instrumentation all have relevance for science teaching and learning. Taking the example of 'reproducibility' reveals that this is an issue that is increasingly important for scientists particularly in the biomedical fields. With the advent of multimedia tools, there is now an emerging body of journals, for example, that are integrating video technology as a component of scientific articles. Take for instance the *Journal of Visualized Experiments*, a peer reviewed journal that describes its mission on its website as follows:

> The Journal of Visualized Experiments (JoVE) was established as a new tool in life science publication and communication, with participation of scientists from leading research institutions. JoVE takes advantage of video technology to capture and transmit the multiple facets and intricacies of life science research. Visualization greatly facilitates the understanding and efficient reproduction of both basic and complex experimental techniques, thereby addressing two of the biggest challenges faced by today's life science research community: i) low transparency and poor reproducibility of biological experiments and ii) time and labor-intensive nature of learning new experimental techniques. (www.jove.com/about)

Given that the use of technological tools in science education is increasing around the world, it is plausible to engage students in similar practices where they can compare, debate and question their experiments captured in video data, creating the opportunity for them to reflect on reproducibility of experimental techniques and scientific data. Comparison and discussion of several experiments generated in the classroom can be conducted to allow students to evaluate how their experiments contribute to collecting reliable data.

## 4.3 Examples of Scientific Practices: Classification, Observation and Experimentation

A necessary companion to the three practices described so far (i.e. observation, classification and experimentation) is representation. According to Suarez (2010), the topic of representation has become a booming topic in philosophy of science as evidenced by the number of conferences, workshops, books and articles produced in the last few years. The topic is at the crossroads of attempts in analytical philosophy to come to terms with the relation between theory and the world, and in the philosophy and history of science to develop a proper understanding of the practice of modeling in the sciences. Scientific representation also overlaps with, and has been claimed to have implications for, metaphysics, the philosophies of mind and language, and aesthetics. Suarez states that the interest from analytical philosophy is related to the notion of reference, and the metaphysics of relations; the interest from philosophy of science is related to an attempt to understand modeling practices. These two distinct forms of inquiry into the nature of representation may be distinguished as the "analytical inquiry" and the "practical inquiry". According to Suarez, although these are types of inquiry that are not mutually exclusive, they impose different demands and point in different directions. Analytical inquiry seems to have historically preceded the practical one, but the relative importance of the latter has grown to the extent that in recent years it has become dominant. This movement takes model building to be the primary form of representational activity (Suarez, 2010), although of course a diversity of representational tools (e.g. figures, graphs, charts, images) are also used in science.

Given the prominence of models in the literature on representation, it is worthwhile to discuss them briefly here. (A more extended discussion on models is provided in Chap. 6). Typically the vehicle of the representation is designated as the 'source'; and the object as its 'target' (Hughes, 1997). Anything can in principle play the role of sources or targets, so these terms are mere place-holders. By contrast, the practical inquiry has avoided questions regarding the nature of the representational relation, focusing instead upon the very diverse range of models and modeling techniques employed in the sciences. The presupposition behind this type of inquiry is that these modeling techniques must be properly understood in their context of application. The literature on modeling in science is immense not just in philosophy of science (e.g. Giere, 1992) but also in science education (e.g. Erduran & Duschl, 2004; Gilbert & Boulter, 2000). Some of the historical key texts include Norman Campbell's (1920) and Mary Hesse's (1966).

The analytical inquiry pursues definition and conceptual analysis, and it emphasizes what Suarez (2010) called the "constitutional question". It is interested in the relation that must conceptually hold between source and target for the source to represent the target. Thus theories of the constituents will typically implicitly answer the question: what is scientific representation? The practical inquiry by contrast focuses on what Suarez calls 'means'. It studies context dependent properties and features of a particular situation that make the source useful for scientists as a representation of the target. It is interested in pragmatic questions regarding the actual workings of models, including judgments of accuracy or faithfulness. Accounts of the means of representation provide case by case analyses of the types of properties, of sources, targets, users, purposes, and context for any given particular representation. A theory of scientific representation needs to address ontological questions such as whether or not or how the representations correspond to what they are meant to represent.

Scientific practices such as observation, classification, experimentation also involve the use of various methods that, result in observational, experimental or historical data sets (see Chap. 5). The different forms of representation that scientists use in obtaining and analyzing data, are intricately embedded in particular cognitive practices involving reasoning that result in modeling, explaining, and predicting. These cognitive practices are coupled with discursive practices that involve argumentation and social certification (see Chap. 7). All of these practices work in concert. Discursive practices do not come at the end of inquiry but they are part and parcel of the conduct of scientific activities, reasoning about data sets, modeling meaningful representations that can be used for explaining and predicting new possibilities. The activities of classifying, observing, or experimenting, individually and collectively, have to result in some level of modeling,[1] explaining, predicting. In that sense, they are not isolated activities (as is sometimes portrayed in science curricula) but rather they are interconnected epistemic practices that work together with discursive practices in an iterative fashion. The outcome of such interactions is the generation of new knowledge that can ultimately be verified through empirical means in the real world.

## 4.4 A Proposed Heuristic of Scientific Practices

The discussion so far presents the case that scientific practices involve particular activities such as observation, classification, and experimentation in a complex set of interactions including collection and analysis of data, and certification of subsequent knowledge claims. While science education reform efforts have often tried to advocate the teaching and learning of a diverse range of processes, products, and mechanisms of science, their coherent presentation from the point of view of students is far from being realized in everyday classrooms. Part of the problem from our point of view is that often these various features of science are taught to learners in a rather disconnected fashion without a sense of an eventual culmination of their aims, roles and functions in science. Consequently, students are likely to leave schooling without having a coherent model of the nature of scientific practices.

There is now a considerable body of work within cognitive science and philosophy of science that argues for the model-based accounts of science (Giere, 1991; Nersessian, 2003). A similar agenda can be extended to science education in terms of generating a model-based approach to the depiction of science for curriculum and instruction purposes. In an analogous spirit, we propose a heuristic that (a) brings together the often disparate components of science (e.g. modeling, social certification),

---

[1] We distinguish between modeling as a scientific practice and models as form of scientific knowledge. Models as a form of scientific knowledge will be discussed in more detail in Chap. 6.

## 4.4 A Proposed Heuristic of Scientific Practices

and (b) redefines the 'discarded' process skills aspects (e.g. experimentation, classification) with the newer practices aspects (e.g. epistemic operations like argumentation and modeling) into one representation that capitalizes on the interrelatedness of scientific practices. It should be noted, however, that the proposed heuristic is not merely grounded in the broad basis of the epistemic, cognitive and social-institutional dimensions of science but aims to communicate a nuanced and holistic interpretation of scientific practices, which can be unpacked relative to its various components. For example, the articulation of social contexts and norms that underlie the social certification of scientific practices can complement the heuristic. Thus the proposed heuristic of scientific practices ought to be broad and comprehensive enough to potentially embrace various interdisciplinary links including links to the economics, politics and history of science which can feed into the understanding of scientific practices.

As discussed elsewhere (Erduran & Mugaloglu, 2013), and revisited briefly here, a model of science could for instance, be characterized from an economics perspective. Radder (2010) distinguished three ideal–typical models of science: Commodified science, autonomous science and public interest science. Commodified science refers to the economic instrumentalization of science. Autonomous and public interest science emphasize criteria other than economics such as development of the society or development of science itself. Autonomous science illustrates an independent scientific community whereas public interest of science frames the function and the role of scientific community with solving or relieving the problems of society. Scientific knowledge as commodity or as "the product of a collective human enterprise to which scientists make individual contributions which are purified and extended by mutual criticism and intellectual co-operation" (Ziman, 1991, p. 3), extends the rational conceptualizations of scientific knowledge to situate it as a product of commerce. Conventional boundaries between scientific endeavor and the societal and cultural norms that surround science could be argued to dissolve through the "science as commodity" idea. Indeed from this perspective, *"science (can) no longer (be) regarded as an autonomous space clearly demarcated from the 'others' of society, culture and economy. Instead, all these domains have become so internally heterogeneous and externally interdependent, even transgressive, that they cease to be distinctive and distinguishable"* (Nowotny, Scott, & Gibbons, 2001, p. 1).

Hence in approximating a heuristic that conveys a range of scientific practices, a systemic approach bringing together the epistemic, cognitive and social-institutional aspects of science is essential for communicating to students a representative account of science. Therefore, we propose a heuristic that targets the often-disparate theoretical accounts of scientific practices and synthesizes them into a whole. The heuristic can be visualized in terms of an analogy with the structure and relationship between components of the benzene ring (Fig. 4.2).

The benzene ring is an organic compound that is composed of six carbon atoms and six hydrogen atoms joined in a ring where one hydrogen atom is attached to each carbon atom. Benzene has a continuous pi bond, which is a covalent chemical bond where lobes of atomic orbitals overlap. In this sense, the pi bonds are diffuse bonds.

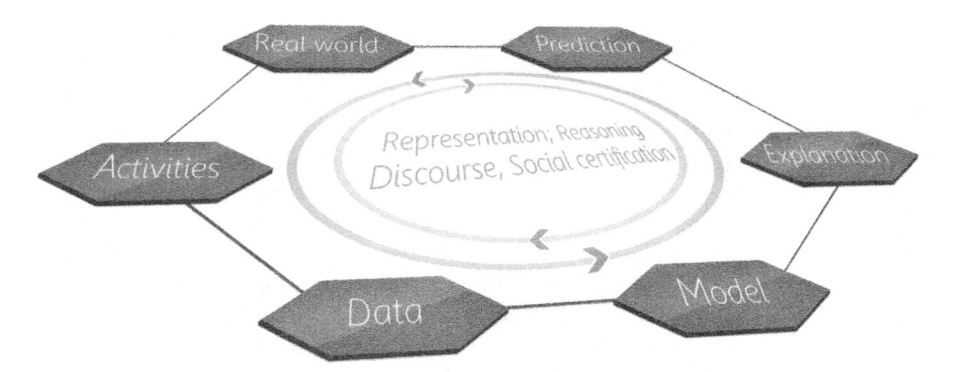

**Fig. 4.2** "Benzene Ring" heuristic of scientific practices

**Table 4.3** Benzene ring analogy

| Analog: Benzene ring | Heuristic: Scientific practices |
| --- | --- |
| Six-carbon hexagonal ring with three double bonds | Each of the carbon atoms in the hexagonal structure represents a scientific practice |
| Double bonds flip around a circle | Scientific practices are not confined to a definitive location in the representation |
| Benzene ring is represented as a hexagon with pi electrons moving around the ring | Representation, reasoning, discourse, social certification and similar processes correspond to the pi electrons. They float around the practices 'ring' but in essence they are integral to and interact with scientific practices |

In this analogy, the various epistemic and cognitive aspects of science are represented as each carbon atom around the ring and the diffuse pi bonds represent the social contexts and practices that apply to all of these aspects. The heuristic illustrates the epistemic and the cognitive dimensions of science as being interrelated and influenced by social dimensions in one holistic representation. The links between the different epistemic components are underlined by the dynamic socio-cognitive processes represented by the electron cloud denoting representation, reasoning, discourse and social certification (among other cognitive, social and institutional factors) which enable the instantiation of each component. The internal ring structure represents the 'cloud' of processes (including the sociological, cultural and economic dimensions) that underlie the epistemic components. The flow is multidirectional and fluid. A significant strength of the heuristic is that the typically disparate science process skills are no longer isolated but are fundamentally redefined and positioned in interactions within and relative to other scientific practices.

Thus, the analogy, further clarified in Table 4.3, communicates to teachers and students that practices of science are interrelated within a range of epistemic, cognitive and social-institutional practices. Overall, the heuristic serves two primary purposes: (a) it illustrates a holistic approach to representing scientific practices, and (b) it provides a pedagogical tool for communicating about scientific practices.

## 4.5 Application of the Benzene Ring Heuristic

The foci of science education curriculum reform efforts across the world are beginning to acknowledge that scientific practices are not peripheral to instruction but are central to science learning. The third wave of science education reform in the USA has advocated a shift from an emphasis on scientific literacy to scientific proficiency (Duschl, Schweingruber, & Shouse, 2007) and from emphasis on scientific inquiry to scientific and engineering practices (NRC, 2012). Meanwhile the science education research community has been witnessing major debates about what constitutes nature of science (Allchin, 2011; Irzik & Nola, 2011; Lederman, 2007; Matthews, 2012). These debates demand a shift in how science educators (a) conceptualise the nature of scientific inquiry, (b) mobilize students' cognitive abilities, and (c) understand practices as located in different interactive spheres of activity (e.g. NRC, 2012). Hence it is worthwhile to ask to what extent the Benzene Ring heuristic has any relevance to curricular policy advocated through the notion of scientific practices within the policy frameworks. Figure 4.3 depicts how scientific

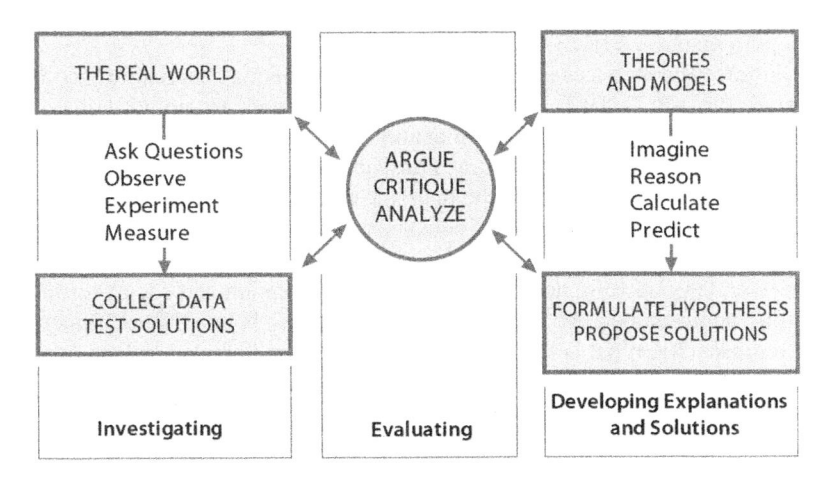

**Fig. 4.3** Scientific practices embedded in the three spheres of activity for scientists and engineers (Reproduced from NRC, 2012, p. 45)

**Table 4.4** Benzene Ring heuristic and NRC (2012) practices

| Feature of Benzene Ring heuristic | NRC practices |
|---|---|
| Real world | ? present in Fig. 4.3 |
| Activities | P1, 2, 3 |
| Data | P4, 5 |
| Model | P2 |
| Explanation | P6 |
| Prediction | ? |
| Reasoning, discourse | P7, 8 |
| Representation, social certification | ? |

and engineering practices are advocated by *A Framework for K-12 Science Education*. Essentially there are eight particular practices (referred to as P1, P2, P3, etc., in Table 4.4) proposed in this document (NRC, 2012, p. 49):

1. Asking questions (for science) and defining problems (for engineering)
2. Developing and using models
3. Planning and carrying out investigations
4. Analyzing and interpreting data
5. Using mathematics and computational thinking
6. Constructing explanations (for science) and designing solutions (for engineering)
7. Engaging in argument from evidence
8. Obtaining, evaluating, and communicating information

The relationships between these practices are further illustrated in Fig. 4.3. The figure shows that the development of theories and models are mediated by a set of activities and norms that involve observation, experiment, critique, analysis and argument to mention a few.

When scientific practices in the proposed "Benzene Ring" heuristic are compared to those in the NRC (2012) document, some similarities are noted. For instance, the notions of practices, models, explanations, and data as well as the reasoning aspects such as critique and analysis are consistent across the Benzene Ring heuristic and the NRC (2012) framework. The heuristic communicates emphasis on the role and significance of representation, social certification and prediction across all scientific practices (Table 4.4).

Thus the Benzene Ring heuristic includes representation and social certification aspects as being embedded in all scientific practices. It is a visual, holistic and iconic representation that is memorable particularly for science teachers given it is based on a science analogy. A significant aspect of this heuristic is that it creates a more dynamic connection between practices embedded in the three spheres of activity represented in Fig. 4.3. Indeed the heuristic bridges non-essential and artificial dichotomies, presenting a more holistic and complex account of scientific practices. Overall the Benzene Ring heuristic consolidates the epistemic, cognitive and social-institutional components of science in a simple, iconic and visual model that is an accessible tool for pedagogical purposes.

The discursive component of the Benzene Ring heuristic in educational settings is easily garnered by the considerable research, for example on argumentation in science classrooms (e.g. Erduran, Simon, & Osborne, 2004; Jimenez-Aleixandre, Bugallo-Rodriguez, & Duschl, 2000; Kelly & Takao, 2002; Sandoval & Reiser, 2004; Zohar & Nemet, 2002); and pedagogical strategies to support writing (e.g. Hand, Prain, Lawrence, & Yore, 1999). Let us take argumentation as an example that can mediate some of the discursive processes that underlie scientific practices. Argumentation is a critically important discourse process in science (Toulmin, 1958). There is now a substantial body of research (e.g. Erduran & Jimenez-Aleixandre, 2008) that has made the case that it should be taught and learned in the science classroom. The learning and teaching of argumentation i.e., the coordination of evidence and theory to support or refute an explanatory conclusion, model or prediction (Suppe, 1998) has emerged as a significant educational goal around the world in recent years. Educating students about how scientists know and what evidence they use to support their claims are critical goals for science education (Erduran & Jimenez-Aleixandre, 2008, 2012). The shift from what-we-know to how-we-know has been argued to require a renewed focus on how science education can promote students' skills in justifying claims with evidence. These claims can center on the practices of observation, classification, and experimentation and support representation. Thus, the various components of the Benzene Ring heuristic can be interrogated from an argumentation perspective, making them accountable to evidence and reason. Sampson and Clark (2006) outlined five criteria that concern the evaluation of knowledge claims: (a) nature and quality of the knowledge claim; (b) how (or if) the claim is justified; (c) if a claim accounts for all available evidence; (d) how (or if) the argument attempts to discount alternatives; (e) how epistemological references are used to coordinate claims and evidence (pp. 659–660). All aspects of scientific practices can embed argumentation as a discourse process that mediates the evaluation of claims made about a particular observation, an experimental procedure or a representation.

## 4.6 Conclusions and Discussion

Science is underpinned by a diverse set of practices that are underpinned by cognitive, epistemic and social-institutional activities. The chapter focused on an example set of practices as well discursive and cognitive processes that permeate scientific practices and proposed a heuristic that could be useful in pedagogical contexts. The features of scientific practices that involve the production of models and explanations, in other words more widely "scientific knowledge," are covered in Chap. 6. A significant aspect of the Benzene Ring heuristic is that (a) it communicates a dynamic set of interactions between the data, models, explanations and predictions that underlie the characterizations of phenomena occurring in the real world, and (b) it integrates the social-institutional and cognitive processes that mediate such interactions through discursive practices like argumentation as well as norms such as social certification, which we unpack in more detail in Chap. 7. We have compared the heuristic to the scientific practices as advocated by the NRC's depiction of scientific practices, illustrating that while some aspects of the NRC practices match some of the components (e.g. practices, data, models, explanations), others (e.g. prediction, real world, representation, social certification) are not explicitly advocated.

The Benzene Ring heuristic has the potential to unify, for teaching and learning purposes, the targeted epistemic, cognitive and social-institutional aspects of scientific practices so that they are implemented in a holistic and coherent fashion at the level of science learning. We are in the process of empirically testing this heuristic in teacher education settings (e.g. Erduran, 2014) to investigate how it can be applied meaningfully in the practical realm of teaching. We anticipate that the heuristic will not only provide a useful analogy given the memorable aspect in relation to the benzene molecule but also that the heuristic will help teachers go beyond some of the limitations of conceptualising science as a step-wise and linear process as is often represented through the conventional scientific method approaches, as we discuss in more detail in Chap. 5. For example, the Benzene Ring heuristic can act as a tool for teachers to design instructional sequences that can build around the various aspects of scientific practices ensuring that by the end of a sequence, a coherent overall picture of science is communicated to students. Another aspect of this heuristic is that it can be potentially used as a tool for interrogating the relationships between its different components and for raising ontological questions about scientific practices and their connection to the "real world". Although we did not focus on such questions in this chapter, we anticipate that the Benzene Ring heuristic can be a useful tool for raising awareness of the ontological commitments of science and scientists.

## References

Ackrill, J. L. (Trans.). (1963). *Aristotle's categories and de interpretatione.* (Translated with notes). Oxford, UK: Oxford University Press.

Allchin, D. (2011). Evaluating knowledge of the nature of (whole) science. *Science Education, 95*(3), 518–542.

Bronowsky, J. (1978). *The origins of knowledge and imagination.* New Haven, CT: Yale University Press.

Brown, J. R. (1991). *The laboratory of the mind: Thought experiments in the natural sciences.* London: Routledge.

Campbell, N. (1920). *Physics: The elements.* Cambridge, UK: Cambridge University Press.

Chinn, C., & Malhotra, B. (2002). Epistemologically authentic inquiry in schools: A theoretical framework for evaluating inquiry tasks. *Science Education, 86,* 175–218.

Davies, R. (1989). The creation of new knowledge by information retrieval and classification. *Journal of Documentation, 45*(4), 273–301.

# References

Dillashaw, F. G., & Okey, J. R. (1980). Test of the integrated science process skills for secondary science students. *Science Education, 64*(5), 601–608.

Duschl, R., Schweingruber, H., & Shouse, A. (2007). *Taking science to school: Learning and teaching in grades K-8*. Washington, DC: National Academies Press (http://www.nap.edu).

Erduran, S. (2007). Breaking the law: Promoting domain-specificity in chemical education in the context of arguing about the periodic law. *Foundations of Chemistry, 9*(3), 247–263.

Erduran, S. (2014). Revisiting the nature of science in science education: Towards a holistic account of science teaching and learning. Plenary lecture. In *Proceedings of the FISER conference, special issue of International Journal of Science and Mathematics Education, ISSN:23-1-251X*.

Erduran, S., & Duschl, R. (2004). Interdisciplinary characterizations of models and the nature of chemical knowledge in the classroom. *Studies in Science Education, 40*, 111–144.

Erduran, S., & Jimenez-Aleixandre, M. P. (Eds.). (2008). *Argumentation in science education. Perspectives from classroom-based research*. Dordrecht, The Netherlands: Springer.

Erduran, S., & Jimenez-Aleixandre, M. P. (2012). Argumentation in science education research: Perspectives from Europe. In D. Jorde & J. Dillon (Eds.), *World of science education: Research in science education in Europe* (pp. 253–289). Rotterdam, The Netherlands: Sense Publishers.

Erduran, S., & Mugaloglu, E. (2013). Interactions of economics of science in science education and implications for science teaching and learning. *Science & Education, 22*(10), 2405–2425.

Erduran, S., Simon, S., & Osborne, J. (2004). TAPping into argumentation: Developments in the use of Toulmin's argument pattern in studying science discourse. *Science Education, 88*(6), 915–933.

Franklin, A. (1986). *The neglect of experiment*. Cambridge, UK: Cambridge University Press.

Giere, R. (1991). *Understanding scientific reasoning* (3rd ed.). Fort Worth, TX: Holt, Rinehart and Winston, Inc.

Giere, R. (1992). *Cognitive models of science* (Minnesota studies in the philosophy of science, Vol. XV). Minneapolis, MN: University of Minnesota Press.

Gilbert, J., & Boulter, C. (2000). *Developing models in science education*. Dordrecht, The Netherlands: Kluwer Academic.

Gooding, D., Pinch, T., & Schaffer, S. (Eds.). (1993). *The uses of experiment: Studies in the natural sciences*. Cambridge, England: Cambridge University Press.

Hand, B., Prain, V., Lawrence, C., & Yore, L. D. (1999). A writing in science framework designed to enhance science literacy. *International Journal of Science Education, 21*, 1021–1035.

Hesse, M. (1966/1962), *Models and analogies in science*. Notre Dame, IN: Notre Dame University Press.

Hjorland, B., Scerri, E., & Dupre, J. (2011). Forum: The philosophy of classification. *Knowledge Organisation, 38*(1), 1–24.

Hodson, D. (1998). *Teaching and learning science*. Toronto, ON: Open University Press.

Hughes, R. (1997). Models and representation. *Philosophy of Science, 64*, S325–S336.

Irzik, G. (2010). Why should philosophers of science pay attention to the commercialization of academic science? In M. Suarez, M. Dorato, & M. Redei (Eds.), *EPSA epistemology and methodology of science launch of the European philosophy of science association* (pp. 129–138). Dordrecht, The Netherlands: Springer. doi:10.1007/978-90-481-3263-8_11.

Irzik, G., & Nola, R. (2011). A family resemblance approach to the nature of science. *Science & Education, 20*, 591–607.

Irzik, G., & Nola, R. (2014). New directions for nature of science research. In M. Matthews (Ed.), *International handbook of research in history, philosophy and science teaching* (pp. 999–1021). Dordrecht, The Netherlands: Springer.

Jimenez-Aleixandre, M. P., & Reigosa, C. (2006). Contextualizing practices across epistemic levels in the chemistry laboratory. *Science Education, 90*(4), 707–733.

Jimenez-Aleixandre, M. P., Rodriguez, A. B., & Duschl, R. A. (2000). "Doing the lesson" or "doing science": Argument in high school genetics. *Science Education, 84*(6), 757–792.

Johnson-Laird, P. (1993). Mental models. In M. Posner (Ed.), *Foundations of cognitive science* (pp. 469–499). Cambridge, MA: MIT Press.

Kelly, G., & Takao, A. (2002). Epistemic levels in argument: An analysis of university oceanography students' use of evidence in writing. *Science Education, 84*(6), 757–792.

Krajcik, J., & Merritt, J. (2012). Engaging students in scientific practices: What does constructing and revising models look like in the science classroom? Understanding a framework for K-12 science education. *The Science Teacher, 79*, 38–41.

Kuhn, T. S. (1977). *The essential tension*. Chicago: University of Chicago Press.

Kwasnik, B. H. (1999). The role of classification in knowledge representation and discovery. *Library Trends, 48*(1), 22–47.

Lederman, N. G. (2007). Nature of science: Past, present, future. In S. Abell & N. Lederman (Eds.), *Handbook of research on science education* (pp. 831–879). Mahwah, NJ: Lawrence Erlbaum.

Latour, B., & Woolgar, S. (1979). *Laboratory life: The social construction of scientific facts*. London: Sage.

Matthews, M. (1994). *Science teaching: The role of history and philosophy of science*. New York: Routledge.

Matthews, M. (2012). Changing the focus: From nature of science (NOS) to features of science (FOS). In M. S. Khine (Ed.), *Advances in nature of science research* (pp. 3–26). Dordrecht, The Netherlands: Springer.

Mayo, D. G. (1996). *Error and the growth of experimental knowledge*. Chicago: University of Chicago Press.

National Research Council. (2012). *A framework for K-12 science education: Practices, crosscutting concepts, and core ideas*. Washington, DC: The National Academies Press.

Nersessian, N. (2003). Abstraction via generic modeling in concept formation in science. *Mind and Society, 3*, 129–154.

NGSS Lead States. (2013). *Next generation science standards: For states by states*. Appendix H. Retrieved from http://www.nextgenscience.org/next-generation-science-standards

Norris, S. (1985). The philosophical basis of observation in science and science education. *Journal of Research in Science Teaching, 22*(9), 817–833.

Novak, J. D., & Cañas, A. J. (2008). *The theory underlying concept maps and how to construct them* (Technical Report IHMC CmapTools 2006–01 Rev 01–2008). Florida Institute for Human and Machine Cognition. Available at: http://cmap.ihmc.us/Publications/ResearchPapers/TheoryUnderlyingConceptMaps.pdf

Nowotny, H., Scott, P., & Gibbons, M. (2001). *Re-thinking science: Knowledge and the public in an age of uncertainty*. Cambridge, UK: Polity Press.

Olson, H. A. (1998). Mapping beyond Dewey's boundaries: Constructing classificatory spaces for marginalised knowledge domains. *Library Trends, 47*(2), 233–254.

Pickering, A. (Ed.). (1992). *Science as practice and culture*. Chicago: University of Chicago Press.

Radder, H. (Ed.). (2003). *The philosophy of scientific experimentation*. Pittsburgh, PA: University of Pittsburgh Press.

Radder, H. (2009). The philosophy of scientific experimentation: A review. *Automated Experimentation, 1*(2), 1–8. doi:10.1186/1759-4499-1-2

Radder, H. (2010). *The commodification of academic research: Analyses, assessment, alternatives*. Pittsburgh, PA: University of Pittsburgh Press.

Rouse, J. (2002). *How scientific practices matter: Reclaiming philosophical naturalism*. Chicago: University of Chicago Press.

Russell, B. (1996/1912). Chapter II: The existence of matter. *The problems of philosophy*. New York: Oxford University Press.

Sampson, V. D., & Clark, D. (2006). Assessment of argument in science education: A critical review of the literature. In S. A. Barab, K. E. Hay, & D. T. Hickey (Eds.), *Proceedings of the seventh international conference of the learning sciences – Making a difference* (pp. 655–661). Mahwah, NJ: Lawrence Erlbaum Associates.

# References

Sandoval, W. A., & Millwood, K. (2005). The quality of students' use of evidence in written scientific explanations. *Cognition and Instruction, 23*(1), 23–55.

Sandoval, W. A., & Reiser, B. J. (2004). Explanation-driven inquiry: Integrating conceptual and epistemic scaffolds for scientific inquiry. *Science Education, 88*, 345–372.

Scerri, E. (2007). *The periodic table: Its story and its significance*. Oxford, UK: Oxford University Press.

Schwab, J. (1962). The concept of the structure of a discipline. *Educational Record, 42*, 197–205.

Shapin, S., & Schaffer, S. (1985). *Leviathan and the air-pump: Hobbes, Boyle, and the experimental life*. Princeton, NJ: Princeton University Press.

Suarez, M. (2010). Scientific representation. *Philosophy Compass, 5*(1), 91–101.

Suppe, F. (1998). Understanding scientific theories: An assessment of developments, 1969–1998. *Philosophy of Science, 67*(Suppl), S102–S115.

Toulmin, S. (1958). *The uses of argument*. Cambridge, UK: Cambridge University Press.

Ziman, J. M. (1991). *Reliable knowledge: An exploration of the grounds for belief in science*. Cambridge, UK: Cambridge University Press.

Zohar, A., & Nemet, F. (2002). Fostering students' knowledge and argumentation skills through dilemmas in human genetics. *Journal of Research in Science Teaching, 39*(1), 35–62.

# Chapter 5
# Methods and Methodological Rules

The myth of the scientific method has received much attention in science education. Yet the discussion of how scientific methods can be reconceptualized for science teaching and learning remains rather limited. In this chapter, scientific methods and methodological rules are articulated to capture their complexity and diversity, and to offer some guidelines for their application in science education. After discussing the limitations of the lock-step scientific method, the case is made for the need to broaden student understanding of the diversity of scientific methods. To this end, a set of heuristics are introduced to represent a range of scientific methods and to illustrate how the evidence derived from them contributes to explanatory consilience. The chapter concludes by illustrating how knowledge about specific methods can be used to reflect on the nature of scientific evidence, and lead to concrete understanding of the role of diverse scientific methods in supporting abstract theoretical claims.

## 5.1 Introduction

Science is an organized activity that is governed by "a number of methods and methodological rules" (Irzik & Nola, 2014, p. 1003). Irzik and Nola (2014) note that methodology is typically accompanied by methodological rules that are heavily discussed by philosophers of science but not given as much emphasis in science education research. These rules includes the following: constructing hypotheses that are testable, avoiding *ad hoc* changes to theories, choosing the theory that is more explanatory, rejecting inconsistent theories, accepting a new theory only if it can explain the successes of its predecessors, using controlled experiments to test causal hypotheses, and using blinded procedures when experimenting on humans subjects. In their earlier work, Irzik and Nola (2011) consider methodological rules to be more of "highly idealized rational reconstructions" than "categorical imperatives". They qualify the sense in which these rules should be understood noting that some of them can be abandoned under certain conditions. They also note that some of the rules have implicit values, and therefore satisfying those values and aims necessitate that rules be followed. For example, using blinded procedures on human subjects has an important value of trustworthiness of findings, so if the aim is to avoid some kind of a preference bias on the part of the experimenters or placebo effect on the part of the research subjects, then the rule has to be followed. Likewise, the social norm of "respect for the environment" (see Chap. 7) can be upheld as an ethical aim or value. This link between methodological rules and scientific aims and values is important in the sense that methodological rules are not arbitrary and they serve broader aims in science (see Chap. 4). Likewise scientific aims and values impact the selection and application of methodological rules.

## 5.2 Beyond the "Scientific Method"

Images of the scientific method can be found in many science textbooks and online sources. The Merriam-Webster online dictionary (n.d.), for example, provides the following definition:

SCIENTIFIC METHOD (noun): principles and procedures for the systematic pursuit of knowledge involving the recognition and formulation of a problem, the collection of data through observation and experiment, and the formulation and testing of hypotheses

The same dictionary defines this term for English Language Learners as follows:

the scientific method technical: the process that is used by scientists for testing ideas and theories by using experiments and careful observation

A slightly different version is presented for children in the following way:

Main Entry: scientific method
    Function: noun
        the rules and procedures for the pursuit of knowledge involving the finding and stating of a problem, the collection of facts through observation and experiment, and the making and testing of ideas that need to be proven right or wrong

These descriptions are intended for the general public and young learners, not for experts. They convey a basic story line about what scientists do. From a science education perspective, the idea of systematic pursuit of knowledge, the emphasis on observational and experimental data, and the testing of ideas is a good place to start – although the language may be too complex for some learners. The notion of proving ideas (as communicated in the definition for children) and the necessity of testing hypothesis are problematic, because they become ingrained in students' conceptions of scientific aims and methods (Dagher & BouJaoude, 2005; Driver, Leach, Millar, & Scott, 1996; Schwartz, 2007).

Discussions of the scientific method in the context of science education seem to follow two distinctive tracks. One track consists of philosophers and historians

## 5.2 Beyond the "Scientific Method"

of science as well as science education researchers who object to the algorithmic and methodologically biased representation of the scientific method. Often these researchers indicate that the scientific method is not a linear process and its representation as such in school science is problematic. Another track consists of many science teachers and textbook writers who use an algorithmic version of the scientific method as a tool to impress on students the idea that scientists use a special method to arrive at scientific knowledge. Here the key justification is that students need simple and cognitively less demanding representation of the methods used in science.

It is useful in this context to distinguish between two senses of scientific method (Halwes, 2000). The first sense refers to a method of disciplined inquiry while the second sense of method refers to a standard method involving a specific procedure. While the first sense is broad in scope and endorses a plethora of methods, the second sense is more restrictive. It is this second sense of the scientific method often communicated in school science that science educators find to be objectionable because it introduces students to a simplistic version of the nature of scientific inquiry. The simplistic elements in this depiction of the scientific method primarily pertain to the linearity of steps, bias towards experimental investigations, and isolation from broader theoretical considerations (i.e. cognitive, epistemic, social, institutional).

Representations of 'the scientific method' in the second sense are not uniform. Woodcock (2013) reports wide variation in the content of scientific method representations that range from 2–3 steps to 11. More typical ones tend to be around five or seven consecutive steps that include the following scientific processes: observing, making a hypothesis, experimenting, analyzing data, confirming or rejecting the hypothesis and making conclusions (see Fig. 5.1). These steps, bundled in a compact package, serve as a handy guide for communicating aspects of scientific inquiry or as a template for structuring reports on science projects.

Woodcock (2013) describes five potentially useful functions for the mythical scientific method: informative, prescriptive, participative, demarcative, and elevative. The informative function is intended "to teach students how science works" (p. 7), while the prescriptive function provides a procedure for doing a school science project. The participative function encourages participation by simplifying a complex process and makes the idea of acting as a scientist accessible to students. The demarcative function assumes that the scientific method provides a distinctive scientific approach to problem solving. The elevative function gives attributes to science a level of rationality and objectivity that are attained from following a 'logical' method. Woodcock describes how each of these functions are misleading and recommends doing away with the label of "scientific method".

Strong opposition to the second sense of scientific method has been voiced repeatedly in the science education as well as the philosophical literature for several decades. This opposition to teaching the lock-step method is based on a number of reasons:

**Fig. 5.1** A popular depiction of the scientific method (GeneseeChemistry, n.d.)

## The Scientific Method

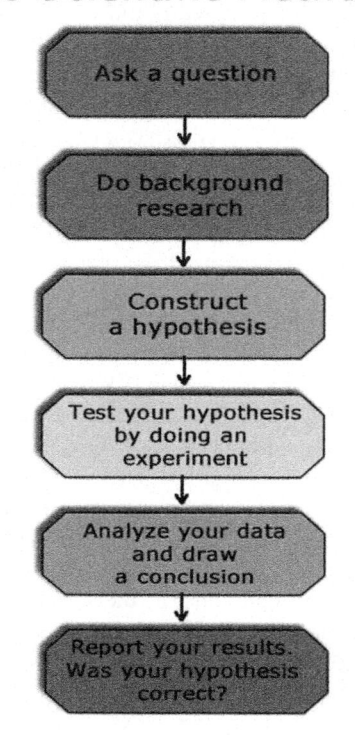

Ask a question
↓
Do background research
↓
Construct a hypothesis
↓
Test your hypothesis by doing an experiment
↓
Analyze your data and draw a conclusion
↓
Report your results. Was your hypothesis correct?

(a) Scientists use multiple methods for investigating questions. When the diversity of methods is overlooked in science education, knowledge obtained from non-experimental methods is likely to be viewed by students and the general public as less privileged or less important than that gained from experimental ones.
(b) When scientists use experimental methods, they do not necessarily follow a rigid algorithm. There are many ways to do experiments, and not all experiments involve hypothesis testing.
(c) Observation as the first step underplays the role of theoretical orientation in focusing the purpose for observations, prior knowledge, and thought experiments in determining what and how to observe.
(d) Repeated emphasis on the scientific method that starts with observations, leads to hypotheses, experiments, and so on contributes to students' rejection of some theories or considering them less scientific because the methods used to collect evidence that supports them did not involve certain components such as forming a hypothesis or conducting an experiment. (Dagher & BouJaoude, 2005)
(e) The scientific method "subverts young learners' understandings of both the practices and the content of the discipline" (Windschitl, Thompson, & Braaten, 2008, p. 942).

## 5.2 Beyond the "Scientific Method"

Calling for a shift from reference to "the scientific method" to "scientific methods" should go beyond fixing the linguistic aspect to address the assumptions and practices that reside at the core of the disputed expression. Woodcock recommends replacing "scientific method" with the "scientists' toolbox" metaphor that he attributes to Wivagg and Allchin (2002). This metaphor focuses on the concept of "scientists as workers who select tools they view as appropriate to solving their problems. This avoids suggesting that scientists follow a canonical recipe" (Woodcock, 2013, p. 9). Windschitl and colleauges (2008) propose "model-based-inquiry" (MBI) as an alternative framework for school science investigations. The MBI approach engages students in interactions regarding what they know and want to know, generating testable hypotheses, seeking evidence and constructing arguments.

Despite strong arguments levied against teaching the scientific method in its lock-step form for several decades, many science textbooks across the globe continue to outline its steps. "The scientific method" has had a resilient presence among practicing teachers. Consensus among science education researchers has not stopped authors from including it in their introductory science textbook chapters, or prevented teachers from emphasizing it. The simplicity of the neatly organized steps seems too appealing to ignore.

One of the shortcomings of promoting "the scientific method" in its popular form is communicating the mistaken notion that there is indeed a "uniform, interdisciplinary method for the practice of good science" (Cleland, 2001, p. 987). The emphasis on "the scientific method" also contributes to the perception that doing credible scientific work necessitates using this method. This leads to the false conclusion that scientists who do not use experimental methods are not likely to arrive at trustworthy knowledge. The consequence of such perception is that historical investigative methods in the natural sciences are often viewed by non-experts to be not as scientific as those that employed experimental methods. Such characterization becomes deeply problematic in cases where it is impossible to conduct experiments, or where experiments contribute only one strand of evidence that is necessary but not sufficient to substantiate claims. This observation supports the need to understand the variety of methodologies used in the sciences which is critical not only for understanding the claims derived from the non-experimental investigations but also for understanding why they constitute valid claims in that particular context and domain.

Reform documents in science education support teaching a variety of scientific methods. For example, the *Next Generation Science Standards* (NGSS Lead States, 2013) dedicates an appendix to a discussion of nature of science, listing eight themes accompanied by a basic outline of target learning outcomes for grades K-2, 3–4, middle school and high school. Two of these themes support the development of scientific practices that are relevant to this discussion: "scientific investigations use a variety of methods" and "scientific knowledge is based on empirical evidence". The content of the learning outcomes for the various grades levels explicitly advocates communicating a pluralistic orientation to scientific methods.

From a philosophical standpoint, Sankey (2008) notes that theorists tend to consist of traditional methodologists who subscribe to "a single, universally applicable method invariant throughout the history of science and the various fields of scientific study" and methodological pluralists who argue for a "plurality of methodological rules governing theory evaluation" (p. 90). According to the second view, science "is not characterized by a single invariant method, but by a set of evaluative rules to which scientists appeal in the context of theory appraisal" (p. 90). However, Sankey argues that methodological pluralism need not result in relativism. He offers five theses that characterize the pluralist account:

(a) Multiple rules: scientists utilize a variety of methodological rules in the evaluation of theories and in rational choice between alternative theories.
(b) Methodological variation: the methodological rules utilized by scientists undergo change and revision in the advance of science.
(c) Conflict of rules: there may be conflict between different methodological rules in applications to particular theories.
(d) Defeasibility: the methodological rules, taken individually rather than as a whole, are defeasible.
(e) Non-algorithmic rationality: rational choice between theories is not governed by an algorithmic decision procedure which selects unique theory from among a pool of competing theories. (Sankey, 2008, p. 92)

These five theses promote the appraisal of scientific theory by "an evolving set of methodological rules" (p. 92). Scientists may apply different weight to these rules that may result in favor of opposing theories.

## 5.3 Scientific Methods and Methodological Rules

In expanding the discussion on scientific methods, types of scientific methods are explored for the purpose of supporting better understanding of scientific practices. Since evidence-based explanations are critical in science, the investigative and analytical methods used to generate evidence are often subjected to scrutiny. Understanding the coordination between claims, evidence and reasoning as key components of scientific explanation, is an important learning goal in science education. Discriminating between acceptable sources of evidence and determining appropriate data-gathering methods demand a nuanced understanding of the object of study and the tools of the trade, so to speak, in a given science domain. The determination of adequate tools and proper methods falls under the purview of domain-specific experts and is not a matter of public opinion. For this reason, it is optimal to reflect on the diversity of scientific methods in the context of learning disciplinary content.

A student in an astronomy lesson would probably have no problem understanding that astronomers do not do experiments on stars and other distant objects, in the sense that they do not manipulate them in space. They do however, make hypotheses and test these hypotheses without having to conduct an experiment per se. When this kind of thinking is shifted to objects within physical reach, the whole picture is transformed. The same student who is now taking a biology class is likely to be more perplexed by the limited role experimental evidence plays in the theory of natural selection. That same student is going to have difficulty with the presented evidence because it is not obvious why appealing to non-experimental data or to

## 5.3 Scientific Methods and Methodological Rules

observational data is necessary when working with objects that are within reach. To be told that there is a historical dimension to the explanation does not make it more understandable. Students need to understand how historical data are obtained, interpreted, and related to other non-historical data sets. They need to overcome a common assumption that testing hypotheses has to involve experimentation. Within philosophy of science, however, testing a hypothesis or a theory involves scrutinizing whether or not observable consequences follow from predictions. Sometimes this sense of 'testing' might be about using a telescope to observe stars, for instance, without doing an experiment. In other instances, an experiment might need to be designed, for instance in the context of the effectiveness of a drug on disease, to check the validity of a prediction. The utility of the Family Resemblance Approach (FRA), as described in Chap. 2 can be illustrated in this scenario. If students were to be taught science taking into account the FRA, then they would begin to scrutinize the domain-general and domain-specific features of various branches of science, including the methodological variations. They would begin to question how one domain is similar and different from another, and indeed what makes a particular domain of science 'scientific' in the first place. They would begin to appreciate the nuances in the way that different sciences operate with methods in particular ways. Science teaching can address the methodological point that data obtained from different sources and different interpretive traditions, across different sciences might converge on making and validating scientific claims, a point that underscores the FRA idea.

Perhaps one of the vexing issues in supporting student understanding of evolutionary theory (ET) stems from the complexity of its structure. That there is no hyper-evolutionary theory as such but rather a system of inter-related theories is important to point out to students. Mayr (2004) specifies five theories: the no-constancy of species, descent of all organisms from common ancestors (branching evolution), gradualness of evolution (no saltations, no discontinuities, the multiplication of species (the origin of diversity), and natural selection. In addition, these theories and the manner in which they are constructed invite interdisciplinary reasoning processes as well as a diverse set of investigations and methods. Some of these methodologies are familiar to students, but some are not. Examples of those methodologies familiar to students are the ones dealing with direct evidence (i.e. observational and experimental). Other methodologies that are less familiar are those dealing with circumstantial and historical evidence (see Fig. 5.2). A person who is less familiar

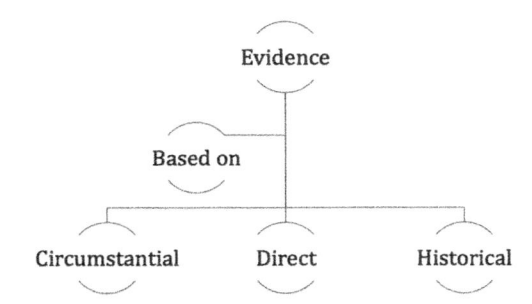

**Fig. 5.2** Nature of evidence obtained in evolutionary theory. Based on Dagher & BouJaoude, (2005, p. 380)

with these methods or who finds them not to be as scientific as the others is likely to conclude that evolutionary theory is more of a metaphysical theory, much in the same way that Carl Popper himself did before he recanted his views in 1978 (Mahner & Bunge, 1997). Evidence that students reject the theory of evolution or claim it to be not as scientific as other theories can be found in empirical studies. Even though religious beliefs or religiosity can play a strong role in motivating students' inclinations, there is also evidence that shows that poor understanding of diverse methods is a factor. In one study for example, a third of students interviewed had a problem with the method of theory generation, arguing that it missed one or more of the steps of the scientific method. Two thirds of the students were concerned about the lack of direct experimentation, and about 13 % were concerned that the theory does not lend itself to making testable predictions (Dagher & BouJaoude, 2005). Gaps in student knowledge of the nature of biological knowledge, how various components of ET connect, and how the various methods contribute to providing a broader explanatory framework are example issues that need to be addressed.

Students' difficulties with ET are further complicated by the difficulties of comprehending the concept of deep time in geology (Catley & Novick, 2009; Dodick & Orion, 2003). Public perceptions of scientific methods tends to echo student perceptions. A Gallup poll (Newport, 2004), reports that in response to the question *"Just your opinion, do you think that Charles Darwin's theory of evolution is _____"*, it was found that

> Just a little more than a third of the American public is willing to agree with the 'scientific theory well supported by evidence' alternative, while the same percentage chooses the 'not well supported by evidence' alternative. Another 30 % indicate that they don't know enough about it to say or have no opinion. There has been essentially no significant change in the responses to this question since 2001. (Newport, 2004)

Knowing the importance of gathering empirical evidence in science is a good starting point, but understanding how this evidence is collected, what methodological rules guide its collection, and how reasoning tools function in validating and justifying explanations demand a much more sophisticated knowledge of science and nature of science. Furthermore, understanding how evidence from a variety of methods leads eventually to evidential consilience (see Fig. 5.3), and subsequent explanatory consilience help situate scientific methods and methodological rules within the larger theoretical structures that they support. The focus shifts from considering the one investigation or method to a family of methods, and to the thought processes and explanatory frameworks that undergird them.

Consilience goes beyond coherence. Wilson (1998) considers consilience to be "the key to unification" (p. 8). He states that "William Whewell, in his 1840 synthesis *The Philosophy of the Inductive Sciences*, was the first to speak of consilience, literally a "jumping together" of knowledge by the linking of facts and fact-based theory across disciplines to create a common groundwork of explanation" (Wilson, p. 8). Explanatory consilience, however, is dependent on evidential consilience and the ability to use findings from different methods or fields to yield explanatory power. A brief description of how the idea of consilience originated and was recently used by Gould and Wilson is found in an essay by Carey (2013).

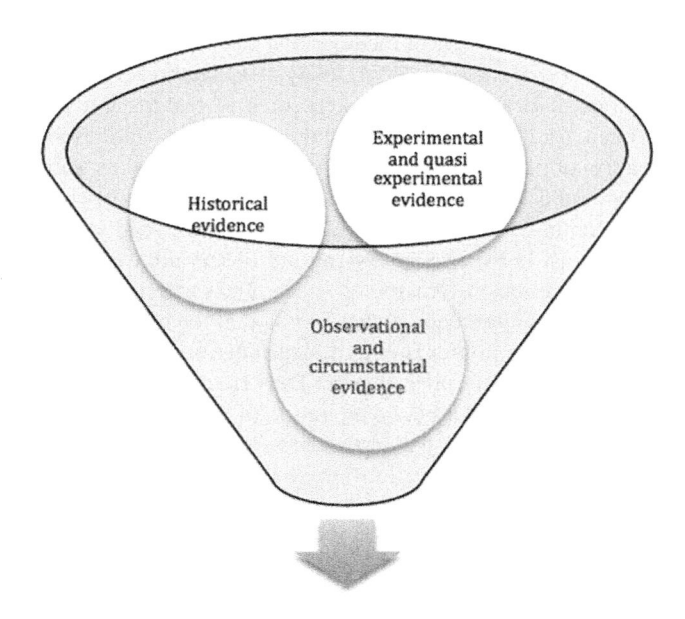

## 5.3 Scientific Methods and Methodological Rules

**Evidential consilience ---> Explanatory consilience**

**Fig. 5.3** Evidential and explanatory consilience through evidence obtained from a variety of methodological sources

What are the varieties of methods scientists use to collect data? In Alverz's (1997) *T. Rex and the Crater of Doom*, the author presents a fascinating chronicle of how experts from different disciplines worked independently and collaboratively to arrive, after decades of problem-solving, at a well-founded, strongly supported explanation for the extinction of the dinosaurs. A variety of experts using different investigative tools and methods rallied to contribute to this question. It would have been impossible to solve the mystery with the knowledge and investigative tools utilized in one science domain. For most intriguing scientific questions, this interdisciplinarity is part of contemporary practices in different branches of science and engineering.

Conveying a sense of interdisciplinarity in the classroom allows teachers to transform the teaching of discrete concepts more towards teaching problem-based investigations that can center on socio-scientific issues or on pure scientific puzzles (such as that of the dinosaurs). At one concrete level, students can be assisted in understanding the variety of scientific methodologies by experiencing them first hand and noting how findings arising from them and from other methods (that they cannot experience first hand) can be marshaled to make broader generalizations, develop and test models, and propose credible explanations. How can this be done? At the level of each content area or set of concepts or problem to be solved, the first step is to determine the parameters of the content and the sets of investigations that can be done in the classroom and those that are addressed by second hand investigations.

In order to elucidate the variety of methods that can be explored in school science, it is helpful to describe Brandon's (1994) analysis of methods used in evolutionary biology. Even though Brandon explores the idea of experiment in one context, his work is pertinent for illuminating ways of thinking about methods in other science domains or in areas of inquiry that require multiple domains, thus utilizing multiple tools and methodologies. Brandon depicts two ways in which experiments are usually contrasted: contrast with observations and contrast with descriptive work. Critical to the contrast between experiment and observation is the occurrence of manipulation that he defines in a restricted sense. The restricted sense of manipulation in this case rules out interventions that do not alter the phenomena. He gives the example of dissection as a non-example of manipulation as it involves the making visible of otherwise invisible phenomena. Thus manipulation in the context of this discussion "involves the deliberate alteration of phenomena" (Brandon, p. 61). An example of manipulation in the restricted sense would be an instance where independent variables are changed to allow the documentation of their effect on dependent variables. (For further discussion on experimentation, see Chap. 4).

In terms of the contrast of experiment with descriptive work, a key factor to the contrast is whether a hypothesis is being tested or whether the values of parameters are being measured. Parameter measures may demand considerable manipulation but may or may not involve the testing of hypotheses. Brandon (1994) gives the following example. If biologists are interested in finding out whether a given herbivore can exert a selective factor for a population of plants, the herbivore (serving as independent variable) would be introduced to an experimental plot and variables pertaining to its effects on plant survival and reproduction, without necessarily posing a hypothesis.

Brandon's examples illustrate that not all experiments involve hypothesis testing and that not all descriptive work is non-manipulative. He represents the connections between experiments and observations in terms of a two-by-two table reproduced in Table 5.1. The nature of the investigation (experiment/observation) is related to whether or not (a) it involves manipulation and (b) hypothesis testing or parameter measure. According to his analysis, one can think in terms of experiment and non-experiments/observations relative to descriptive versus experimental methods.

In considering how evidence gathered from multiple types of investigations work together, it is possible to show the role of different methods in the production of findings, eventually contributing to explanations in a given domain. In the case of evolutionary biology, specifically, the theory of evolution through natural selection, we can use the matrix in Table 5.1 to insert examples (see underlined text in

**Table 5.1** Types of observational and experimental methods

|  |  | Experiment/observation | |
|---|---|---|---|
|  |  | Manipulate | Not manipulate |
| Descriptive/ experimental | Test hypothesis | Manipulative hypothesis test | Non-manipulative hypothesis test |
|  | Measure parameter | Manipulative description or measure | Non-manipulative description or measure |

## 5.3 Scientific Methods and Methodological Rules

**Table 5.2** Observational and experimental methods in the context of the theory of evolution through natural selection

|  | Manipulate | Not manipulate |
| --- | --- | --- |
| **Test hypothesis** | Manipulative hypothesis test | Non-manipulative hypothesis test |
|  | *e.g. Investigations in Genetics-molecular evolution* | *e.g. Observation of Darwin's finches* |
| **Measure parameter** | Manipulative description or measure | Non-manipulative description or measure |
|  | *e.g. Artificial selection and breeding* | *e.g. Studies in paleontology and developmental biology* |

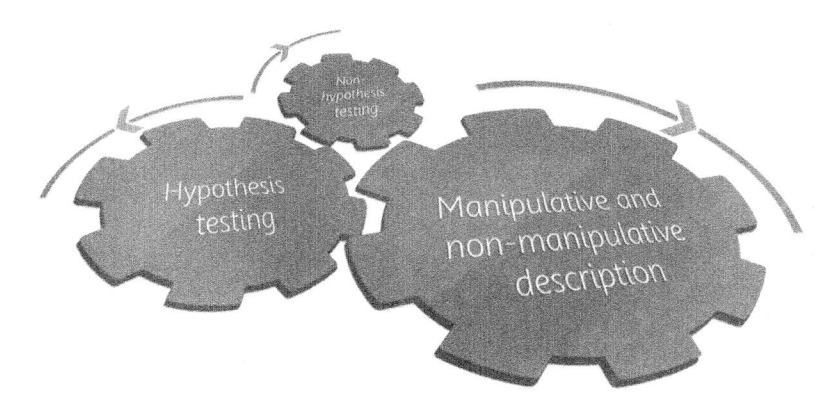

**Fig. 5.4** The 'gears' image illustrating how evidence from a variety of methods works synergistically to contribute to explanatory consilience

Table 5.2) to illustrate how substantial evidence for the theory arises from the consideration of findings obtained through multiple methods.

The convergence of evidence from different methods can then be used to lead to a broad explanatory structure. It is not one method or one line of experimental or observational evidence that support complex theoretical claims but several lines of evidence need to be synthesized to bring about the level of theoretical rigor that is typically associated with established scientific knowledge. Components of evidence from these different sources become gears, so to speak, that drive the 'engine' of explanatory consilience (Fig. 5.4).

The preceding discussion can be used as a way of thinking about varieties of scientific methods typically taught in school science. The two-by-two table allows the depiction of the range of methods used in different domains and how some domains may emphasize some methods more than others. The size of the gears of in Fig. 5.4 can be reallocated to correspond to where the bulk of the evidence comes from.

Let us consider an example from chemistry as illustrated in Table 5.3. In a review of the Periodic Table, Eric Scerri describes how Mendeleev predicted the existence

**Table 5.3** Observational and experimental methods in the context of periodicity of elements

|  | Manipulate | Not manipulate |
|---|---|---|
| **Test hypothesis** | Manipulative hypothesis test | Non-manipulative hypothesis test |
|  | *e.g. Crookes' study of gases* | *e.g. De Boisbaudran's discovery of gallium* |
| **Measure parameter** | Manipulative description or measure | Non-manipulative description or measure |
|  | *e.g. Rutherford's artificial transmutation of elements* | *e.g. Mendeleev's prediction of gallium* |

of the element gallium (or eka-aluminum) through a non-manipulative description coupled with quantitative reasoning about atomic weights:

> Mendeleev could interpolate many of the properties of his predicted elements by considering the properties of the elements on each side of the missing element and hypothesizing that the properties of the middle element would be intermediate between its two neighbors. Sometimes he took the average of all flanking elements, one on each side and those above and below the predicted element. This interpolation in two directions was the method he used to calculate the atomic weights of the elements occupying gaps in his table, at least in principle. (Scerri, 2007, p. 132)

Scerri further states that it was the French chemist Emile Lecoq De Boisbaudran who subsequently "worked independently by empirical means, in ignorance of Mendeleev's prediction, and proceeded to characterize the new element spectroscopically" (Scerri, 2007, p. 135). De Boisbaudran was testing the hypothesis of the existence of a new element by spectral analysis of an ore and managed to isolate gallium through this method. The manipulative aspect of some chemical methods include (a) Crookes' study of gases where pressure and voltage were used as variables in spectroscopical study elements (e.g. Scerri, p. 251) as an example of manipulative hypothesis testing, and (b) Rutherford's artificial transmutation of elements through bombardment of nuclei with protons (e.g. Scerri, p. 253) as an example of manipulative description. All together these methods, along with numerous others, contributed to the collective and eventual depiction of elements (Table 5.3).

Identifying features of investigations that distinguish them as belonging to one of the four quadrants does not necessarily mean that all investigations in any one quadrant are carried out in the same way, or follow an algorithm. All it means is that these methods share some distinctive features (e.g. involve hypothesis testing or conducting observations). Thus the diversity of methods within any of the four genres is preserved. It is important then to point out that we are not advocating the replacement of "the scientific method" with "four scientific methods". Rather, the point of referring to the matrix is to emphasize the range of ways in which investigations can be set up to address different research questions. Furthermore, the use of heuristics such as the two-by-two tables can provide meta-tools for communicating how different science domains or particular examples from within a domain of science might be employing different methods. Seen in this way, the heuristic reinforces the Family Resemblance Approach in the characterization of science (in particular scientific methodology, in this case), specifying in the preceding

## 5.3 Scientific Methods and Methodological Rules

examples, how the evolutionary theory in biology and periodicity in chemistry contrast in terms of methodological approaches.

From an educational perspective, what is important about the concepts represented in this chapter's tables and figures is not only where the investigations and eventually evidence comes from, but also where they lead to; how each methodological approach has to be validated on its own; and how methods fit in with a robust web of explanations. It is such aspects that are missing in school science, and are in need of stronger emphasis in science teaching and learning. More emphasis on the intricate nature of how methodologies relate to the formation of theories, laws and models would provide a sense of purpose to doing investigations and direction for making sense of findings. It would also help students see the connections between the visible components, for instance, the apparatus in an experiment, and the larger purpose that scientists engaged in these investigations aspire to find out.

If we further focus on one particular methodology, we see that even a particular approach can be unpacked into its various dimensions. For example, Brandon (1994) points out that the idea of manipulation is one of degree. Pointing to the case of quantum physics, he acknowledges that any degree of observation effects a change in the observed phenomenon, making the lower right hand part of Table 5.1 non-applicable. Also it could be argued that it is possible to use the same data to test a hypothesis and to measure parameters. Sometimes hypotheses get formed after performing parameter measurement. The purpose of the experiment and the way in which it is carried out determine whether it falls under hypothesis testing or parameter measurement. Splitting hairs over where a given investigation fits within these categories can be instructive but is not necessary from a pedagogical standpoint.

While methods can be conceptualized as dichotomies, Brandon (1994) states that it is profitable to view them as components of two continua that range from testing to not testing and from manipulation to non-manipulation. A given branch of science can utilize a continuum of methods. He represents this relationship in the way depicted in Fig. 5.5 whereby investigations can be viewed as more (upper left corner) or less (lower right corner) experimental.

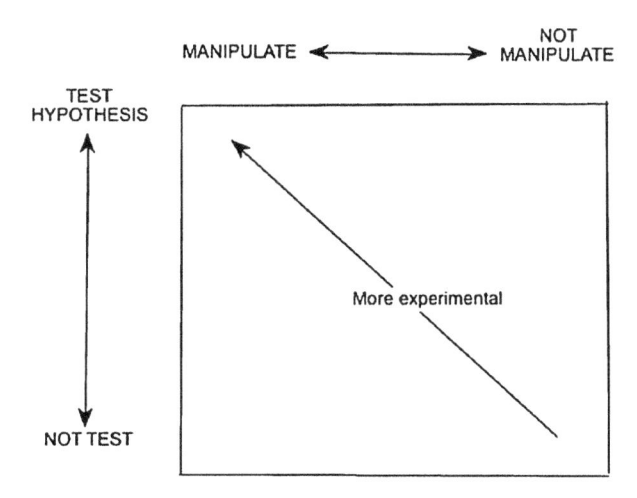

**Fig. 5.5** Brandon's representation of the "space of experimentality" between two continua (Reproduced from Brandon, 1994, p. 66)

One aspect of scientific methodology that we did not consider so far involves thought experiments which are widely used by scientists in solving problems. Thought experiments, typically "performed in the laboratory of the mind" (Brown, 2011), are considered by Nersessian to be "a species of *simulative model-based reasoning*, the cognitive basis of which is the capacity for mental modeling" (Nersessian, 1991, 1992 cited from 2008) and do not fit well into a specific category or side of the chart. Rather, they are reasoning tools that can be accommodated somewhere along the continuum model depending on whether they lead to testable claims. As Nersessian states, "real-world and thought experimental narratives aim at enabling the expected audience to reason through either carrying out an experiment or executing it imaginatively" (Nersessian, 2008, p. 173).

The preceding discussion illustrates that scientists use diverse investigative methods that can be classified at a basic level into one of four categories each of which can be subsequently unpacked for further scrutiny. Scientists can be also involved in investigations that do no fit squarely into one of these classes of methods but fall somewhere on the continuum between non-manipulation to manipulation and non-testing to testing. Finally, in many cases, a combination of these investigative methods is typically used by scientists to answer complex questions within and across disciplinary boundaries.

## 5.4 Methodological Rules as Evolving Entities

Earlier in this chapter, a few methodological rules were mentioned such as constructing hypotheses that are testable, using controlled experiments to test causal hypotheses, and using blinded procedures when experimenting on human subjects. Discussion of methodological rules in the philosophical literature addresses a range of issues that relate scientific aims to scientific methods. This discussion is not without controversy. For example, Resnik (1993) argues against connecting methodological rules to scientific aims because aims are too broad to attend to methodological specifics.

Methodological rules vary dramatically in dimension and scope (or grain-size). In the ethical dimension for example, methodological rules are expected to consider and take into account the rights of human subjects. At the macro-level of theory choice, it is possible to talk about methodological rules that are different in scope from the practical rule of conducting controlled experiments for testing causal hypotheses. At the micro-level of performing investigations, methodological rules can be established to minimize a variety of error types. Allchin (2012) offers a taxonomy of errors that includes the following error types: material, observational, conceptual and derived that span from the immediate and local to the derived and global contexts. Knowledge of these potential error types can result in methodological rules that, if implemented, can help minimize them. Table 5.4 provides examples of error types sampled from Allchin.

## 5.5 Educational Implications

**Table 5.4** Taxonomy of error types sampled from Allchin (2012, p. 912)

| Error type | Error |
|---|---|
| Local ↕ Derived (Global) | **Material**<br>–Improper materials<br>–Failure to differentiate similar phenomenon through controlled conditions |
| | **Observational**<br>–Insufficient controls to establish domain of data or observations<br>–Incomplete theory of observation |
| | **Conceptual**<br>–Flaw in reasoning (includes simple computational error)<br>–Inappropriate statistical model |
| | **Discursive**<br>–Communication failures: incomplete reporting<br>–Mistaken credibility judgments (Matthew effect, halo effect) |

Allchin (2012) advocates using historical cases to learn about the tentativeness of scientific knowledge and the role of uncovering errors at different levels to improve knowledge. Attention to potential sources of error, in the context of developing or validating methodologies is needed as a way of establishing methodological rules in K-12 science classrooms. Error types that are closer to the local end of the spectrum lend themselves to discussion and application in the course of ongoing classroom investigations. Awareness of these errors is significant for understanding the historical institution of some methodological rules. Using examples from Allchin's work (see Table 1 in Allchin, p. 906), it is possible to see how avoiding a placebo effect can be done through a method that employs blind clinical trials, an observer effect by a double-blind method, coincident variables through controlled experiments, and sampling errors through statistical analyses and so on. In turn, this awareness can be revived in the classroom through discussion of historical cases, current scientific research, or actual classroom investigations.

## 5.5 Educational Implications

Supporting the development of sophisticated understanding of scientific methodology in science education requires educating students about the variety of questions that could be posed and the different ways of investigating them. Far from lecturing students about scientific methods, the aim should be to involve them in the practice of designing investigations and understanding methodological options that scientists use to address relevant questions. Appendix H of *The Next Generation Science Standards* focusing on nature of science lists two main ideas pertaining to scientific methods and the learning outcomes associated with them across K-12 schooling (see Table 5.5). The contents of Table 5.5 supports the discussion outlined in the previous section. Students should be involved with multiple investigations that

**Table 5.5** Two categories pertaining to scientific methods and corresponding understandings about NOS across K-12 science education in Appendix H of the *Next Generation Science Education Standards* (NGSS Lead States, 2013, p. 5)

Understandings about the nature of science

| | K-2 | 3–5 | Middle school | High school |
|---|---|---|---|---|
| Scientific investigations use a variety of methods | Science investigations begin with a question | Science methods are determined by questions | Science investigations use a variety of methods and tools to make measurements and observations | Science investigations use diverse methods and do not always use the same set of procedures to obtain data |
| | Scientist use different ways to study the world | Science investigations use a variety of methods, tools, and techniques | Science investigations are guided by a set of values to ensure accuracy of measurements, observations, and objectivity of findings | New technologies advance scientific knowledge |
| | | | Science depends on evaluating proposed explanations | Scientific inquiry is characterized by a common set of values that include: logical thinking, precision, open-mindedness, objectivity, skepticism, replicability of results, and honest and ethical reporting of findings |
| | | | Scientific values function as criteria in distinguishing between science and non-science | The discourse practices of science are organized around disciplinary domains that share exemplars for making decisions regarding the values, instruments, methods, models, and evidence to adopt and use |
| | | | | Scientific investigations use a variety of methods, tools, and techniques to revise and produce new knowledge |

## 5.5 Educational Implications

| Scientific knowledge is based on empirical evidence | Scientists look for patterns and order when making observations about the world | Science findings are based on recognizing patterns | Science knowledge is based upon logical and conceptual connections between evidence and explanations | Science knowledge is based on empirical evidence |
| --- | --- | --- | --- | --- |
| | | Scientists use tools and technologies to make accurate measurements and observations | Science disciplines share common rules of obtaining and evaluating empirical evidence | Science disciplines share common rules of evidence used to evaluate explanations about natural systems |
| | | | | Science includes the process of coordinating patterns of evidence with current theory |
| | | | | Science arguments are strengthened by multiple lines of evidence supporting a single explanation |

**Table 5.6** Anticipated outcomes when giving students opportunities to engage in reasoning about a diverse set of scientific methods

| Students ask | Students learn to |
| --- | --- |
| What method(s) do I use to investigate or test this claim? | Consider a variety of methods and determine how to investigate claims |
| What makes this method more effective than an alternative one in investigating my question? | Reflect on alternative methods and consider their potential in contributing to knowledge |
| How does this method/finding contribute to a broad explanation? | Go beyond the mechanics of the investigation and the immediate result, and aim for explanatory synthesis |

expose them to the range of methodological diversity involved in the science domain they are studying. Furthermore, explicit reference to different types and sources of evidence is important for helping students appreciate how they contribute to the formulation of coherent explanations.

The science curriculum can support the development of better understanding of scientific methods by including more explicit reference to these methods. Instructional practices can build on this by including specific meta-cognitive strategies and heuristics. One way of promoting such development is by holding discussions with students about the best ways to approach the questions they are about to investigate. Teachers can use existing resources (such as NAS, 2008 in the case of evolutionary theory) in conjunction with Table 5.1 to compare and discuss the variety of methods used by scientists or those proposed by students. The Table can help structure the discussion regarding the feasibility of these methods in the classroom and their utility in scientific contexts. The main purpose is to create awareness of methodological choices and their role in developing different forms of scientific knowledge such as theories, laws and models. It can support teachers in promoting a view of science as a "set of practices [that] shows that theory development, reasoning, and testing are components of a larger ensemble of activities" (NRC, 2012, p. 43). In addition, this "focus on practices (in the plural) avoids the mistaken impression that there is one distinctive approach common to all science—a single 'scientific method'" and serves to illustrate that "practicing scientists employ a broad spectrum of methods" that fit the purpose of the research (NRC, p. 44). When students are provided with the opportunity to reason about a diversity of methods, they are more likely to develop a more sophisticated understanding of how and why scientific methods function in science (see Table 5.6).

To put these ideas into action, teachers can play the following roles:

(a) Invite students to develop questions and design investigations to provide explanations that explain target phenomena;
(b) Focus students' attention on the relationship of the methods they use in their designs to their claims, and to their conclusions;
(c) Support students in understanding that choosing investigative methods is question dependent (and feasibility dependent as well). Not all investigations need a hypothesis or involve hypothesis testing, or doing experiments;
(d) Identify how a given method answers a limited set of questions;

(e) Compare, when applicable, the methods students use to those used by scientists;
(f) Reflect on the utility of a combination of scientific methods in the field they are studying for providing evidential and explanatory consilience (as pointed out in Fig. 5.4);
(g) Conduct discussions to reflect on key empirical methods that scientists used to produce the evidence underlying major theories. Keeping track of these on classroom charts and revisiting them as the school year progresses support a deeper level of understanding of these key ideas.

The scheme borrowed from Brandon (1994) shown in Table 5.1 can be used in conjunction with the 'gears' image presented in Fig. 5.4 to:

– Illustrate the diversity of methods in specific science domains;
– Demonstrate how these methods relate to one another or contribute to a broader conceptual scheme;
– Problematize the complexity of methods e.g. not all hypothesis testing is manipulative (contrary vision promoted by school science);
– Bring coherence to how different methods are related within any one domain of science and how these are used to develop explanations and theories;
– Serve as an efficient meta-cognitive tool for reflecting with students on the diversity of scientific methods and reflecting on the explanations gained from findings;
– Illustrate how empirical findings obtained from more than one type of method (represented by a quadrant) are necessary for validating the theoretical structures, or explanatory narrative;
– Go beyond the methods per se to reveal a broader purpose whether it pertains to generating/verifying explanations or synthesizing new products (not only intellectual but also actual artifacts/materials);
– Make the point that there is nothing inherently superior about any of these methods. The selection of method is judged in terms of its potential for informing the question and the nature of the discipline.

Providing learners with opportunities to experience a variety of scientific methods that generate different types of evidence leading to more coherent explanations allows them to gain a better understanding of the complexity of scientific work. Among the ideas worth teaching is that it is not one investigation but several that make or break a scientific theory. This entails giving students opportunities to reflect on how empirical evidence sought from a variety of sources through a range of methods leads to justified explanations. The 'gears' in Fig. 5.4 can be used as a visual tool in reinforcing the idea that evidence can be generated through different methods and used to different extent in establishing explanatory frameworks in science.

As discussed earlier in this chapter, students' experience with scientific methods consist of following prescribed procedures. Students are seldom asked to engage in designing investigations or validating methods. Even those fortunate enough to be given opportunities to plan their investigations, may not get a full understanding of the methodological options they may want to consider in relation to pursuing their question. Engaging students in planning and reflecting on the methods they use brings them closer to authentic scientific inquiry and ensures that they are actively engaged in the processes of methods construction, evaluation, validation and revision. Moreover, students need to understand that the methods scientists employ in the field, as well as the ones they use in the classroom, are not unchangeable recipes to follow. They are constructed to achieve specific goals. They are limited by technological tools, guided by methodological rules, subject to modification and refinement, and evolve over time.

The images relayed through observational instruments or data obtained from experiments are often taken for granted by learners. For this reason, it is worthwhile to call student attention to the fact that the scientific instruments have changed over time and have evolved to more sophisticated types that permit making observations or testing hypotheses about complex phenomena in ways that were not possible before. This can promote a better appreciation of the possibilities afforded by improved instrumentation and the intricate relationship between science and technology. Overall, articulating the range of methods used in science will facilitate students' appropriation of a broader range of scientific practices. Taking the field of evolutionary biology, for example, students can be asked to organize the variety of observational, experimental, and historical methods they are studying visually in a matrix of scientific methods that documents the variety of evidence each of these methods present, and trace the contribution of evidence from a variety of sources to broader theoretical claims. In educational settings that focus science learning over a long period of time on single science field, the matrix can be used to discuss theories within each of these fields (e.g. ecology, geology, physics, and chemistry).

In educational settings involving general science courses that include topics from different sciences, the matrix shown in Table 5.1 can become an effective way to revisit how scientific methodologies used across the disciplines differ in emphasis but not necessarily in the rigor of the theoretical claims resulting from them. In addition, the role of instruments that are commonly used in observational studies that students cover in lessons (e.g. microscope in biology, telescope in astronomy) can be discussed (a) to connect their work to instruments used by early scientists (using brief anecdotes), and (b) to compare the data obtained by the instruments used in the classroom to data obtained by scientists.

The pluralistic nature of scientific methods must be reflected in curriculum materials and teaching practices. The cost of not doing so is that students are left confused and the prospects of participating in the sciences are limited. It is indeed questionable if students are participating in science at all when they are immersed in contexts where they are following recipes mindlessly without (a) reflecting on the practices of scientists in developing scientific knowledge through particular methods, and (b) being engaged in the design, evaluation and execution of authentic methods themselves. One critical aspect of scientific methodology to be discussed with students is the revisionary nature of the scientific methods themselves. Scientific methods should be viewed as creative aspects of scientific knowledge growth that can be discussed, questioned and contested just as the claims that arise from data collected through the implementation of these methods. Viewing scientific methods as efficient tools that lie at the foundation of knowledge construction allows them to be regarded as revisable and subject to change. Thus, it is important for students to engage in the construction, evaluation, validation and revision of methods to understand their "fit for purpose" in achieving particular knowledge claims.

## 5.6 Conclusions

Four representations capture the spirit of this chapter: Brandon's matrix (Table 5.1) and "Space of experimentation" (Fig. 5.5), relationship between evidential and explanatory consilience (Fig. 5.3), and synergy of evidence gears (Fig. 5.4). The tables and figures used to clarify several ideas in this chapter can serve a dual role as conceptual tools and pedagogical heuristics. They can be used to guide curriculum planning and instruction, and they can be used as instructional tools to organize discussions on scientific methods with students in the context of teaching specific topics. The goals for these discussions, ideally conducted in relevant domain-specific contexts, are twofold: (a) to enhance student understanding of the diversity of scientific methods, and (b) to engender student appreciation of the contribution of these methods to the construction of theories, laws and models. From our standpoint, the first constitutes the proximate goal and the second constitutes the ultimate goal. For this reason, we specifically choose Fig. 5.4 (synergy of evidence gears) to emphasize as a key contribution of this chapter. The epistemic content, diversity and depth associated with scientific methods and methodological rules could and should be more explicitly expressed in school science. This does not require spending considerable instructional time on these issues. Rather, it demands mindful infusion of these goals in science lessons. It demands engaging students in the design, use, evaluation and reflection on scientific methods and methodological rules in the construction of scientific knowledge.

## References

Allchin, D. (2012). Teaching the nature of science through scientific errors. *Science Education, 96*, 904–926.
Alvarez, W. (1997). *T. Rex and the crater of doom*. Princeton, NJ: Princeton University Press.
Brandon, R. (1994). Theory and experiment in evolutionary biology. *Synthese, 99*, 59–73.
Brown, J. (2011). *The laboratory of the mind: Thought experiments in the natural sciences*. New York, NY: Routledge.
Carey, T.V. (2013, March/April). Consilience. *Philosophy Today*. Retrieved April 30, 2013, from http://philosophynow.org/
Catley, K., & Novick, L. (2009). Digging deep: Exploring college students' knowledge of macroevolutionary time. *Journal of Research in Science Teaching, 46*(3), 311–332.
Cleland, C. (2001). Historical science, experimental science, and the scientific method. *Geology, 29*, 987–990.

Dagher, Z., & BouJaoude, S. (2005). Students' perceptions of the nature of evolutionary theory. *Science Education, 89*, 378–391.

Dodick, J., & Orion, N. (2003). Measuring student understanding of geological time. *Science Education, 87*(5), 708–731.

Driver, R., Leach, J., Millar, R., & Scott, P. (1996). *Young people's images of science*. Buckingham, UK: Open University Press.

GeneseeChemistry. (n.d.). *Scientific method*. Retrieved from http://geneseechemistry.wikispaces.com/Week+1+-+Scientific+Method

Halwes, T. (2000). *The myth of the magical scientific method*. Retrieved from http://www.dharma-haven.org/science/myth-of-scientific-method.htm

Irzik, G., & Nola, R. (2011). A family resemblance approach to the nature of science. *Science & Education, 20*, 591–607.

Irzik, G., & Nola, R. (2014). New directions for nature of science research. In M. Matthews (Ed.), *International handbook of research in history, philosophy and science teaching* (pp. 999–1021). Dordrecht, The Netherlands: Springer.

Mahner, M., & Bunge, M. (1997). *Foundations of biophilosophy*. Berlin, Germany: Springer.

Mayr, E. (2004). *What makes biology unique?* Cambridge, UK: Cambridge University Press.

*Merriam-Webster Dictionary* (n.d). Retrieved May 24, 2013, from http://www.merriam-webster.com/dictionary/scientific%20method

National Academy of Sciences. (2008). *Science, evolution and creationism*. Washington, DC: National Academies Press.

National Research Council. (2012). *A framework for k-12 science education*. Washington, DC: National Academies Press.

Nersessian, N. (2008). *Creating scientific concepts*. Cambridge, MA: MIT Press.

Newport, F. (2004). *Third of Americans say evidence has supported Darwin's evolution theory*. Retrieved June 1, 2013, from http://www.gallup.com/poll/14107/third-americans-say-evidence-has-supported-darwins-evolution-theory.aspx

NGSS Lead States. (2013). *Next generation science standards: For states, by states*. Appendix H. Retrieved from http://www.nextgenscience.org/next-generation-science-standards

Resnik, D. (1993). Do scientific aims justify methodological rules? *Erkenntnis, 38*, 223–232.

Sankey, H. (2008). *Scientific realism and the rationality of science*. Aldershot, UK: Ashgate.

Scerri, E. (2007). *The periodic table: Its story and its significance*. Oxford, UK: Oxford University Press.

Schwartz, R. (2007). What's in a word? *Science Scope, 31*(2), 42–47.

Wilson, E. O. (1998). *Consilience: The unity of knowledge*. New York: Alfred A. Knopf.

Windschitl, M., Thompson, J., & Braaten, M. (2008). Beyond the scientific method: Model-based inquiry as a new paradigm of preference for school science investigations. *Science Education, 92*, 941–967.

Wivagg, D., & Allchin, D. (2002). The dogma of "the" scientific method. *The American Biology Teacher, 69*(9), 645–646.

Woodcock, B. (2013, June 19–22). *"The scientific method" on trial*. Paper presented at the International History and Philosophy in Science Teaching biennial meeting, Pittsburgh, PA. http://archive.ihpst.net/2013-pittsburgh/conference-proceedings/

# Chapter 6
# Scientific Knowledge

This chapter describes the various forms of scientific knowledge such as theories, laws and models emphasizing their coherence and contribution to the growth of scientific knowledge, a theme that tends to be invisible in school science. The discussion is guided by the following questions: How are the forms of scientific knowledge related? What functions and roles do they play in the development of scientific knowledge? Are there disciplinary variations in theories, models, laws and explanations in science? What example features of scientific knowledge are important to promote in science teaching and learning? The chapter begins by articulating the idea that theories, laws and models (TLM) work together in generating scientific explanations that lead to knowledge growth. This is followed by an account of select epistemological issues that are relevant for science education. These include levels and kinds of theories, domain-specificity of scientific laws, nature of models and explanatory pluralism. Taken together, these themes illustrate significant nuances about scientific knowledge including disciplinary similarities (in the sense of the Family Resemblance Approach) as well domain-specificity in different branches of science such as chemistry, biology and physics.

## 6.1 Introduction

Theories, laws and models (TLM) are forms of scientific knowledge that work together to generate and/or validate new knowledge. They are products of the scientific enterprise. For example, the atomic theory, the periodic law of elements and molecular models all contribute to understanding of the structure of matter (Table 6.1). In science, as theories, laws, and models evolve they amass a broader empirical and/or mathematical base, and they lead to additional understandings akin to Thomas Kuhn's notion of science as a puzzle solving activity (Kuhn, 1970).

**Table 6.1** Theories-Laws-Models (TLM) in different science domains

| Domain form of knowledge | Biology | Chemistry | Physics |
|---|---|---|---|
| Theory | Genetic theory | Atomic theory | Thermodynamics |
| Law | Inheritance law | Periodic law | Laws of thermodynamics |
| Model | Genes | Atomic model | Heat transfer |
| TLM *explain* | Biological traits | Structure of matter | Heat |

The theoretical structures that form the foundations of a given paradigm grow, and get applied to broader contexts. For example, the atomic theory gets expressed and used in various domains of science including not only physics and chemistry but also, for instance, molecular biology (e.g. molecular model of DNA). If at some point a drastic paradigm shift occurs, this gives rise to new cycles of knowledge growth independently of the previous one or engulfs it. History of science is full of examples of paradigm shifts, for instance Lavoisier's theory of chemical reactions and combustion versus the phlogiston theory in chemistry, and Mendelian inheritance versus pangenesis in biology.

Although school science is dominated by theories, laws and models, often characterised as "content knowledge", there is little in the way of building students' understanding of how various forms of scientific knowledge relate to each other, and how they contribute to scientific explanations in a given scientific discipline in a specific topic. Indeed, it is very rare that school science represents growth of scientific knowledge in terms of theories, laws and models working together as a coherent and dynamic system that result in understanding how the universe and the natural world function. What would be the goal of presenting in school science scientific knowledge and its growth as a dynamic system? What would students understand about the nature of science in this fashion that is not currently or conventionally included in instruction? Understanding the mechanisms of knowledge growth would ensure that students distinguish scientific knowledge as a coherent network of theories, laws, and models, rather than as discrete and unrelated pieces of information. A holistic and relational representation and presentation of scientific knowledge is likely to promote students' meaning making of why and how we know what we know in science. Simply presenting these forms of knowledge in a textbook or instruction does not guarantee that the logic of the relationship between them is obvious to the learner.

Consider a representation of theories, laws and models operating within a particular research tradition as coherent and interrelated sets of knowledge forms that enable scientific understanding through emergent explanations about the universe. The abbreviation TLM in the rest of this chapter is used to indicate reference to this intricate network of relationships between these forms of knowledge. Thus TLM does not refer to the individual components of the knowledge forms but to the interconnected components in a given context. Explanation underpins the relation-

## 6.1 Introduction

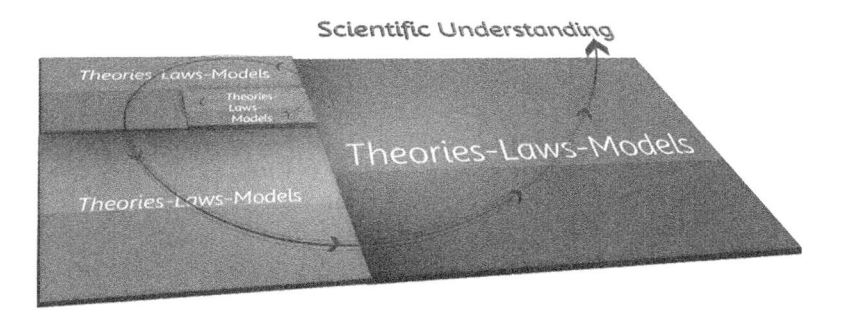

**Fig. 6.1** TLM, growth of scientific knowledge and scientific understanding

ships among the TLM components. Theories, laws and models should provide coherent explanations that collectively lead to scientific understanding (see Sect. 6.5). Figure 6.1 illustrates how TLM work together at a point in time to explain particular phenomena. As scientists learn more about phenomena, the TLM get enriched and scientific knowledge grows leading to increased scientific understanding. For example, in the context of biology, genetic theory, inheritance laws and the concept of the gene work together to explain biological traits of living organisms. Each plane of knowledge growth could be considered as a paradigm which can at times be abandoned and replaced by new sets of TLM, accounting for not only Kuhn's (1962/1970) 'normal' but also 'revolutionary' phases of science.

The depiction of TLM in a knowledge growth framework addresses several issues that are often missing in school science. The TLM framework brings together often disparate pieces of knowledge forms and makes them coherent. Teachers and students can be supported with a visual representation on growth of scientific knowledge as a dynamic and coherent system. Although earlier calls have been made for providing teachers with growth of scientific knowledge frameworks as pedagogical tools (Duschl & Erduran, 1996), science curriculum materials and teaching continues to ignore this important metacognitive dimension in addressing scientific knowledge. The TLM is consistent with the Family Resemblance Approach to science as described in Chap. 2, because it highlights how different disciplines of science resort to a cohesive set of theories, laws and models to explain phenomena that concern their disciplinary questions. TLM also respects the particular structures of the disciplines because even though all sciences rely on particular theories, laws and models, the precise nature of these forms of knowledge will be domain-specific. As we will illustrate later in this chapter, for instance, there may be domain-specific conceptualisations of laws. Science disciplines share conceptual expression of their main knowledge claims through TLM in a domain-general sense, while exercising a nuanced domain-specific articulation. Since particular theories, laws and models in different branches of science offer disciplinary variations, the depiction of scientific knowledge in this fashion opens up the discussion to consider domain-specific features of scientific knowledge. In summary, while there is an overall coherence among TLM components of scientific knowledge in terms of explaining and predicting phenomena, the ways in which different sciences articulate each individual TLM component may vary.

One key concept addressed by science educators engaged in Nature of Science (NOS) research has been the articulation of the differences between a scientific theory and a scientific law (e.g. McComas, 1998). A growth of scientific knowledge framework goes beyond such 'atomistic' differentiations and focuses on a whole set of relationships between different forms of scientific knowledge in unison. Such holistic consideration of theories, laws and models in relation to (as opposed to contrast between) one another is likely to assist learners in understanding the specific features of each knowledge form as well as how they function with one another to explain and predict phenomena. Emphasis on growth of scientific knowledge in terms of a coherent set of theories, laws and models raises new questions for science educators that can be pursued through research. For instance, in what ways can we articulate the relationships between theories, laws and models in contexts that are accessible and meaningful for learners? What are some of the disciplinary variations in TLM in science and what implications do they have for the design of science curricula?

The chapter focuses on select aspects of TLM to represent some important epistemological themes that underpin the growth of scientific knowledge. For example, theories and models come in various categories which illustrate taxonomies of scientific knowledge. Knowledge growth itself is mediated by a set of mechanisms including the evaluation of models through a set of criteria. Overall, TLM presents a stronger 'package' of knowledge forms because it not only includes the knowledge forms themselves but also highlights the dynamics of knowledge growth through a set of heuristics, criteria and standards that drive the epistemic practices of science. The TLM framework thus offers a dynamic system of knowledge forms that illustrate the explanatory power of scientific knowledge in a coherent fashion. The historical and developmental aspects of knowledge growth can be represented and the domain general as well as domain specific elements of scientific knowledge can be interrogated within this framework.

## 6.2 Classification of Scientific Knowledge Forms

The nature of theories, laws and models is often not discussed at a metal-level in school science. As a consequence, students face difficulties in understanding the various levels and classifications of various forms of scientific knowledge. This section will review some relevant characteristics of theories, models and laws.

Consider how theories are treated in schooling. Theories are often confused as guesses, and much of their significance in science is lost on students. For example, there are established theories, and there are fringe theories. A clear understanding of theories that involves some differentiation between levels would help students distinguish established theories from fringe theories and realize that the theory of evo-

## 6.2 Classification of Scientific Knowledge Forms

lution for instance, is one of the established theories of science. Understanding the distinctions between the different theories within a science topic and across topics and disciplines, can help students appreciate not only the significance of theory but its empirical, logical and mathematical bases.

Duschl (1990) refers to levels of theories that are derived from the work of Dutch (1982), who proposed that scientific theories can be classifiable to centre, frontier or fringe regions of science. This conceptualization of types of theories is similar to that proposed by Lakatos (1978), who argued that theories can be assigned to either the hard or soft core of a discipline. Theories at the center level are established as part of mainstream science. Examples include the theory of relativity, the Newton's laws of motion and Kepler's laws of planetary motion. The frontier level is also part of mainstream science, yet there are still unresolved aspects. In other words, these theories are still being challenged by rival explanations. An example would be the cold fusion theory, which may or may not be elevated to a center level theory in due course. Duschl gives evolution as an example of the frontier level of scientific theories. The amount of evidence in science now has elevated evolutionary theory to one of the center level theories of science. Likewise, the extinction of dinosaurs has moved forward from the fringe level at this time. What these examples illustrate is that theories are indeed open to evaluation and that science progresses through evaluation and revision of theories. The central and frontier theories constitute the hard core of science. They are the key set of assumptions and standards on which scientists base their knowledge at a particular timeframe. The fringe level is the point when theories are beginning to make entry into science. At this level, theories may be based on crank ideas or they could be sound, but in either case they are subjected to analysis. In time, they may, or may not, rise to the other two levels.

The wider context of theories in science possesses a long history in philosophy of science as well science education. The conceptual change theories of learning have relied heavily on the characterizations of students' learning as being analogous to theory development in science. For example, students' understanding of the atomic theory might go through the stages of development ranging from Dalton's atomic theory to the quantum mechanical depictions of the atom. Furthermore, science and school science are full of examples where students are asked to contemplate competing theories. For example, some topics include plate tectonics, big bang/steady state, geocentric/heliocentric, and phylogiston/oxygen. It is worthwhile to note, however, that the status of theories within philosophy of science is not straightforward, for instance in relation to models. As Aduriz-Bravo (2013) indicates:

> Logical positivists and critical rationalists established a long intellectual tradition that placed scientific theories at the centre of philosophical analysis, transforming them into the 'structural and functional units' of the science building. For these first 'professional' philosophers of science, theories had logical priority over models; models were considered derivative entities, hierarchically subordinate. These philosophers studied what they saw as 'successful' theories (e.g., Newtonian mechanics) or 'failed' theories (e.g., the theory of phlogiston) without paying much attention to the nature of scientific models. The underlying assumption was that there existed in fact some real entities that satisfied the constraints

imposed by the linguistic propositions constituting a theory—such entities being the models of the theory. (Aduriz-Bravo, 2013, p. 1598)

Aduriz-Bravo (2013) further differentiates three major conceptions of models that link models to theories. First, within the logical positivism (from 1920 to 1960), a scientific model was any example of a theory. However the theory was deemed to be the key object of epistemological analysis. From 1950 to 1980 within the "new philosophy of science," the model is a paradigmatic example of a theory. A subsequent period from 1970 to 2010 emphasised the semantic conception of scientific theories.

Models have been extensively studied in separate lines of inquiry in cognitive psychology and philosophy of science as well as science education (Chi, Feltovich, & Glaser, 1981; Clement, 1989; Gable & Bunce, 1984; Glynn & Duit, 1995). The vast amount of literature on models offer numerous categories and taxonomies of this form of scientific knowledge. Just as theories can be classified with respect to different levels of development, models can be classified from a range of perspectives including epistemic, cognitive and educational perspectives. Models are instrumental in summarizing data, visualizing invisible structures and processes, making predictions, justifying outcomes and facilitating communication in science. A model is typically defined as a representation between a source and a target (Duit & Glynn, 1996; Grosslight, Unger, Jay, & Smith, 1991; Justi, 2000), the target being an unknown object or phenomenon to be explained and the source being a familiar object or phenomenon that helps to understand the target.

Philosophers of science often situate models as intermediaries between the abstractions of theory and the practical actions of experiment (Redhead, 1980). They examine explanatory power of models (Cartwright, 1983; Woody, 1995) and the relation of models to theories (Giere, 1991). Cognitive psychologists study the role of models in cognitive development (Johnston-Laird, 1983; Rogers & Rutherford, 1992) and individuals' model-based reasoning in specific domains such as physics and mathematics (Schauble, Glaser, Duschl, Schulze, & John, 1994). While cognitive psychologists refer to models from the perspective of personal and subjective mental models, anthropologists emphasise the cultural and intersubjective aspects of models (D'Andrade, 1992; Geertz, 1973; Shore, 1996). Depending on the purpose, however, models can be described in various ways (Mihram, 1972) which are detailed below.

Bruner (1966) identified three types of models: enactive, iconic and symbolic or conceptual. An *enactive* model refers to the way in which an individual can translate his or her experience into a model of the world through action. For example, scientists might mimic a phenomenon with their hands in trying to solve a problem, such as the modeling of collision being by fists hitting each other. Bruner's second category is the *iconic* model, which is based on summarizing images. These are physical representations of their prototypes. Examples include maps and small scale buildings that are constructed to provide data for the design of the full-scale versions. It is common practice to identify a 'model' with 'prototype', something to serve as a standard or to be copied. Kuhn (1962/1970), for example, used the word 'model' in this sense when he discussed networks of achievements that scientific communities acknowledge as providing the foundations for further practice. Bruner's third type of model, *symbolic* or *conceptual* model, is a mental construct that may range from the simple descriptive to the rigorously analytical, and in which the symbolism may be as varied, for instance, a pattern of thought or an algebraic equation.

Giere (1991) proposed three types of models: scale, analog and theoretical. Giere's notion of *scale* models is similar to Bruner's iconic models in the sense that scale models share similarity of structure with real objects. *Analog* models involve development of a theory of a new system based on similarities it shares with a known system. Analog models can be illustrated by the early attempts to develop a theory of atom from an analogy with the solar system. Giere's third category is *theoretical* models: a system primarily based on language. Theoretical models are created by formulating and arranging statements in order to define a system. As an example, a Newtonian particle system is a theoretical model that consists of the three laws of motion and the law of universal gravitation.

From an educational perspective, Gilbert and Boulter (1997) differentiated between mental, expressed, consensus and teaching models. *Mental* model is a cognitive representation of an event, object or a phenomenon. *Expressed* model is that version of a mental model that is expressed by an individual through action, speech or writing. *Consensus* model is an expressed model subjected to testing by a social group. The social group can be the scientific or classroom communities who have agreed that a model has some merit relative to some criteria. In the sense that a consensus model is a model negotiated within a community, it is regarded as an extension and modification of personal mental models. A *teaching* model is a specially constructed model used to aid the understanding of a consensus model.

It is possible to consider models from various levels of organisation. Boulter and Gilbert (1996) proposed classification of models based on typologies. Typologies suggest particular models which are representative types and which exemplify groupings. Typologies – taxonomies and partonomies – are derived from psychological accounts on the classification of objects and phenomena at large (Tversky, 1989). Boulter and Gilbert suggest that partonomies of models are based on functional and structural aspects of models, and taxonomies are based on subordinate and superordinate levels in a hierarchical arrangement of models. Boulter and Gilbert (1996) propose three typologies of models: primary taxonomy, performance partonomy and exemplary taxonomy. *Primary taxonomies* emphasize material or symbolic and static or dynamic features of models. *Performance partonomies* are based on certain aspects of models such as structure or behavior. The focus here is on the particular parts of models that allow the analysis of structural and functional aspects of models. *Exemplary taxonomies* group students' expressed models, which are models that emerge through the curriculum or teaching models.

Laws, like theories and models, also offer a rich ground where categories of knowledge can be articulated. Some of the domain-specific categorizations and characterizations of laws will be highlighted in a later section. The key message in the discussion here is that scientific knowledge in the form of TLM comes in taxonomies. Calling attention to the different categories and senses of theories, laws and models helps remove confusion about what these knowledge forms are, how they are different from one another, and how diverse they are. A focus on taxonomies of scientific knowledge can further help formulate comparisons between classroom models, teachers' models and students' models of particular scientific phenomena. Although we are not advocating that these taxonomies be taught as such in school science, they could provide some useful guidelines for how meta-perspectives on categorization of TLM can be considered for teaching and learning purposes. For example, teachers could elicit in classroom discussions the levels of theories and types of models to help overcome conventional misconceptions such as theories as being mere guesses. These points of clarification should arise naturally at opportune moments rather than be the center of the discussion. When theories, laws and models are covered in instruction, it is not always apparent to students that they are forms of scientific knowledge and that they are related to one another. Mere coverage of theories, laws and models as discrete entities in science lessons does not guarantee that students will understand how they collectively and coherently contribute to explanatory and predictive frameworks in science.

The implications of the preceding discussion are that (a) there are classifications of TLM (e.g. levels of theories and types of models), which need to be differentiated and covered in science teaching and learning, (b) the various classifications of theories, laws and models can contribute to students' understanding of scientific knowledge as a coherent system of knowledge forms, and (c) instructional approaches are to ensure that students can be supported in understanding TLM as a coherent unit of scientific knowledge which can be specific in different branches of science. The last point about specificity of TLM in particular scientific domains is an aspect that is often overlooked in science lessons. Students are rarely encouraged to ask questions such as "Are laws in chemistry and physics the same? If not, how are they different?" and "Do theories in physics and biology have the same characteristics?". The next section draws thus attention to the issue of domain-specificity of scientific knowledge.

## 6.3 Domain-Specificity of Scientific Knowledge

TLM are instantiated in different domains of science in various ways. Even though all branches of science develop and are based on theories, laws and models (ie. Family Resemblance of TLM), their precise nature can be rather different in each domain (ie. domain-specificity). As an example, the focus of this section is on laws. Scientific laws are a tricky form of scientific knowledge in educational settings because they are often defined as observed or mathematical regularities. Students often think that they are purely derived from inductive observations, and that they express rules about how nature works. They also think that laws are the end-points of science, that they are confirmed theories (Dagher, Brickhouse, Shipman, & Letts,, 2004; McComas, 1998). Furthermore, it is not clear that students know that laws of nature explain and predict. Indeed, the very nature of laws is not clearly articulated

## 6.3 Domain-Specificity of Scientific Knowledge

in school science. Some laws can be expressed in algebraic form (e.g. Newton's laws of gravitation) while others are qualitative approximations (e.g. Mendeleev's periodicity). Some are probabilistic (e.g. gas laws) while others are definitive (e.g. Avogadro's law).

Volumes have been written about the nature of laws in science. Perspectives on the purpose and nature of laws have changed over time, and some aspects of them continue to undergo some debate. There are various senses of 'laws' in different branches of science as discussed elsewhere (Dagher & Erduran, 2014) some of which are revisited in this section. A law is typically defined as a regularity. But what distinguishes a law of nature from any other regularity? Traditional definitions of a scientific law typically refer to "a true, absolute and unchanging relationship among interacting elements" (Dhar & Giuliani, 2010, p. 7). This traditional view has been challenged on various grounds. Lange (2005) argues that the condition of truth alone does support this distinction since other regularities are also true. He proposes the following four criteria to aid in distinguishing laws of nature from other regularities: necessity, counterfactuals, explanatory power, and inductive confirmations.

Mahner and Bunge (1997) have argued that some laws are said to be "spatially and temporally boundless", where other laws may be "bounded in space and time." Cartwright's critique of "the limited scope of applicability of physical laws" (Ruphy, 2003) problematizes the "truth" aspect of laws. Giere (1999) on the other hand holds the view that what has come to be known as "laws of nature" are in fact historical fossils, holdovers from conceptualisations first proposed in the Enlightenment. He proposes the consideration of models, which he argues are more reflective of how science is actually *practised*. Attempting a balanced description of scientific laws is a complex undertaking considering debates among philosophers about criteria invoked to distinguish between various types of laws such as strict versus *ceteris paribus* laws, empirical versus *a priori* laws. Such criteria include mathematization, necessity, and the potential for explanatory and predictive power. Some of these debates will be revisited in the context of the specific sciences later in this chapter.

Predating recent debates about laws, Bunge (1961) classification of lawlike statements from various philosophical standpoints into more than seven-dozen kinds led him to call for less stringent philosophical restrictions regarding what could be classified as a law:

> There are as many classifications of law statements as viewpoints can be profitably adopted in their regard, and there seems to be no reason—save certain philosophical traditions—why most law statements should be regarded as nonlaw statements merely because they fail to comply with either certainty, or strict universality, or causality, or simplicity, or any other requisite found necessary in the past, where science seemed to concern itself exclusively.... That lawlike (a posteriori and general in some respect) statements be required corroboration and systematicity in order to be ranked as law statements, seems to fit contemporary usage in the sciences. (Bunge, 1961, p. 281)

Bunge's pragmatic view is especially poignant from both a philosophical and pedagogical standpoint. This is because contemporary debates about what counts as a scientific law, argued with core propositions of particular science disciplines, seem to be fundamentally grounded in normative or pragmatic standpoints. Perhaps the most valuable context for such debates has been relative to the role of laws in generating or supporting scientific explanations (Press, 2009).

Domain-specific characterizations provide useful frameworks for understanding what exactly a law in biology, chemistry, or physics (to name few overarching disciplines) does. Until the emergence of philosophy of chemistry as a formalized area of study, the status of laws in chemistry has received little attention within philosophy of science (e.g. Cartwright, 1983). With the upsurge of philosophy of chemistry in the 1990s, there has been more focus on what might make laws distinctly chemical in nature. Some philosophers of chemistry (e.g. Christie & Christie, 2000) as well as chemical educators (e.g. Erduran, 2007) have argued that there are particular aspects of laws in chemistry that differentiate them from laws in other branches of science with implications for teaching and learning in the science classroom. A topic of particular centrality and relevance for chemical education is the notion of "Periodic Law," which is not typically characterised as such:

> Too often, at least in the English speaking countries, Mendeleev's work is presented in terms of the Periodic Table, and little or no mention is made of the periodic law. This leads too easily to the view (a false view, we would submit), that the Periodic Table is a sort of taxonomic scheme: a scheme that was very useful for nineteenth century chemists, but had no theoretical grounding until quantum mechanics, and notions of electronic structure came along. (Christie & Christie, 2003, p. 170)

Some laws in chemistry like the Avogadro's Law (i.e. Equal volumes of gases under identical temperature and pressure conditions will contain equal numbers of particles) are quantitative in nature while others are not. For example, laws of stoichiometry are quantitative in nature and count as laws in a strong sense. Others rely more on approximations and are difficult to specify in an algebraic fashion. As a key contributor to philosophy of chemistry, Eric Scerri takes the position that some laws of chemistry are fundamentally different from laws in physics (Scerri, 2000). While the emphasis in physics is on mathematization, some chemistry laws take on an approximate nature:

> The periodic law of the elements, for example, differs from typical laws in physics in that the recurrence of elements after certain intervals is only approximate. In addition, the repeat period varies as one progresses through the periodic system. These features do not render the periodic law any less lawlike, but they do suggest that the nature of laws may differ from one area of science to another. (Scerri, 2000, p. 523)

The periodic system may not appear law-like from the perspective of a physicist (Scerri & McIntyre, 1997). Significantly, the periodic law seems not to be exact in the same sense that characterise most laws of physics, for instance Newton's laws of motion. The Periodic Law states that there exists a periodicity in the properties of the elements governed by certain intervals within their sequence arranged according to their atomic numbers. As Scerri and McIntyre (1997) discuss, the crucial feature that distinguishes this form of 'law' from those found in physics is that chemical periodicity is approximate. For example, the elements sodium and potassium represent a repetition of the element lithium, which lies at the head of Group I of the periodic table, but these three elements are not identical. Indeed, a vast amount of chemical knowledge is gathered by studying patterns of variation that occur within vertical columns or groups in the periodic table. Predictions which are made from the so called Periodic Law do not follow deductively from a theory in the same way in which idealized predictions flow almost inevitably from physical laws, together with the assumption of certain initial conditions.

Scerri (2000) contrasts the nature of laws in physics such as Newton's Laws of Gravitation and laws in chemistry such as the Periodic Law. Even though both the Periodic Law and Newton's Laws of Gravitation have had success in terms of their predictive power, the Periodic Law is not axiomatized in mathematical terms in the way that Newton's Laws are. Part of the difference has to do with what concerns chemists versus physicists. Chemists are interested in documenting some of the trends in the chemical properties of elements in the periodic system that cannot be predicted even from accounts that are available through contributions of quantum mechanics to chemistry. Christie and Christie (2000), on the other hand, argue that the laws of chemistry are fundamentally different from the laws of physics because they describe fundamentally different kinds of physical systems. For instance, Newton's Laws described above are strict statements about the world, which are universally true. However the Periodic Law consists of many exceptions in terms of the regularities demonstrated in the properties and behaviors of elements. For the chemist there is a certain idealization about how, elements will behave under particular conditions. In contrast to Scerri (2000), and Christie and Christie (2000), Vihalemm (2003) argues that all laws need to be treated homogeneously because all laws are idealizations regardless of whether or not they can be axiomatized. van Brakel further questions the assumptions about the criteria for establishing 'laws':

> If one applies "strict" criteria, there are no chemical laws. That much is obvious. The standard assumption has been that there are strict laws in physics, but that assumption is possibly mistaken . . . Perhaps chemistry may yet provide a more realistic illustration of an empirical science than physics has hitherto done. (van Brakel, 2000, p. 141)

Christie and Christie (2000) indicate that taking physics as a paradigmatic science, philosophers have established a set of criteria for a "law statement," which "had to be a proposition that (a) was universally quantified, (b) was true, (c) was contingent, and (d) contained only non-local empirical predicates" (p. 35). These authors further argue that such a physics-based account is too narrow and applies only to simple systems. More complex empirical sciences do not necessarily conform to such accounts of laws:

> The peculiar character of chemical laws and theories is not specific to chemistry. Interesting parallels may be found with laws and theories in other branches of science that deal with complex systems and that stand in similar relations to physics as does chemistry. Materials science, geophysics, and meteorology are examples of such fields. (Christie & Christie, 2000, p. 36)

The debates around the nature of laws in chemistry are ongoing, and it is beyond the scope of this chapter to capture their full complexity. However, it is important to recognize that philosophers of chemistry continue to dispute the nature of chemical knowledge at large and the nature of laws in particular.

In summary, the suggestion offered by Christie (1994) is considered useful:

> Ultimately the best policy is to define 'laws of nature' in such a way as to include most or all of the very diverse dicta that scientists have chosen to regard as laws of their various branches of science. If this is done, we will find that there is not a particular character that one can associate with a law of nature. (Christie, 1994, p. 629)

So what are the implications of this philosophical discussion for science education? The discussion, focused on chemistry, instantiates how scientists in a given domain articulate laws that have a set of characteristics that may be somewhat different than laws articulated in another domain. A law can be expressed in an algebraic form or it can be approximate and expressed qualitatively within the same science discipline. Specifying the nature of laws, their similarities and differences across domains of science would have implications for science education. For example, textbooks should elicit the approximate nature of the Periodic Law and specify the reference to the patterns in periodicity as an instance of law while highlighting how this law differs from other laws with which the students are familiar. The juxtaposition of the empirical versus theoretical dimensions of orbital models should be differentiated to clarify the different epistemological status of the Periodic Table in light of its historical and empirical foundation. Furthermore, Erduran (2007, p. 257) proposed that an argumentation framework could offer a useful pedagogical strategy for eliciting different characterisations of laws, and suggested a potential activity could be structured as follows:

> [Claim] 1: The periodic law and the law of gravitation are similar in nature. The term "law" can be used with the same meaning for both of them.
> [Claim] 2: The periodic law and the law of gravitation are different in nature. The term "law" cannot be used with the same meaning for both of them.

These claims could be presented with evidence that would support either claim, both or neither. For example, the statement "a law is a generalization" could support both claims while "the periodic law cannot be expressed as an algebraic formula while the law of gravitation can be" could support the second claim. The task for the students would be to argue for either claim and justify their reasoning. Further statements can be developed that would act as evidence for either, both or neither claim. The inclusion of a debate on the nature of laws in a comparative context between physics and chemistry will carry into the classroom the kind of thinking that helps students question the function of laws and consider what laws enable us to do. Without a sense of a debate, textbooks and teachers tend to reinforce the "received view" of science that tends to gloss over the details of what a law is and project a perception of a consensus when there is none. Furthermore, there is a missed golden opportunity to promote understanding of laws not just as generic forms of scientific knowledge but also in terms of their distinctions in different branches of science. In summary, the inclusion of meta-perspectives offered by philosophical accounts of laws can provide useful information in defining laws for science education purposes.

## 6.4 Evaluation of Scientific Knowledge

Scientific knowledge grows through the development, extension and revision of theories, laws and models that explain and predict phenomena in the universe. However the ways in which TLM get established depends on numerous standards of the scientific community. For example, scientists evaluate new evidence in relation to existing models and decide if and how such evidence contributes or not to existing accounts. At times when such evaluations become increasingly unsatisfactory, the entire paradigm subsumed in TLM can be revised. The criteria that drive evaluation of TLM is important to understand as they help illustrate how TLM come to be in the first place and how they contribute to scientific knowledge growth.

Let's take the example of models to illustrate why evaluation criteria might be significant in educational settings. As illustrated earlier, there is a wide range of models which present an additional need to understand their nature (Erduran & Duschl, 2004). The treatment of models in school science tends to be rather simplistic. Teachers typically use physical, pictorial, or scale models to convey how a micro or macro object looks like. But it is not clear that they explore with students how to think with models as conceptual tools, and what precisely these models do. Helping students distinguish physical models from conceptual models is an important learning goal. The practice of modeling involves much more than developing a physical representation. The actual task of learning about or developing a scientific model, which is a form of scientific knowledge, is affected by what teachers and students think a model is. Therefore identifying the purpose of the task and the kind of model needed and what it is needed for (i.e., its function) can help focus a particular activity on deeper content structures.

In school science, a popular activity intended for students to use and communicate models in science involves the use of an 'edible cell model' to illustrate the structure and organization of the cell. Yet, a pizza or candy cell model (e.g. http://www.llemonade.com/cell.html) using gummy worms, raisins and twizzlers to represent different cell organelles, is neither an example of modeling as scientific practice nor a model in the sense of a scientific tool. It is a visual kinesthetic re-representation of a cell model. It is static and does not lend itself to the explanatory or predictive aspect of scientific models. It is perhaps a limited pedagogical tool to communicate visual elements of a scientific concept. It does not address the epistemic and functional aspects of models as forms of scientific knowledge. Consider a contrasting scenario where the students themselves start modeling the components of an onion cell through a set of investigations in the classroom. These might involve observations through a microscope and getting students to derive representations to explain what each component does, how it is related to another and in what ways the representations may or may not capture the actual cells. For example, students could be tasked to identify different shapes and question how their form might relate to their function. The purpose here would be to get the students to generate preliminary criteria for classifying organelles on the basis of structure and function. A teacher who is aware of the different kinds of models and the purpose they serve in a science domain is more likely to be more selective of the models they share and more attentive to the kinds of "modeling practices" they ask their students to engage in. Scientific models that are simplified for pedagogical reasons need to be shared, presented, co-constructed as explanatory or predictive tools that further understanding of key ideas and mechanisms. The learning outcomes in both classes are very different.

What constitute modeling practices? How do models get evaluated and established? Andrea Woody (1995), a philosopher of science, identified four properties of models: approximate, projectability, compositionality and visual representation. These properties could potentially act as useful criteria for guiding and evaluating modeling practices. A model's structure is *approximate*. In other words, the model is an approximation of a complete theoretical representation for a phenomenon. The model omits many details based on judgments and criteria driving its construction. Another characteristic of a model proposed by Woody is that a model is *productive* or *projectable*, meaning that a model does not come with well defined or fixed boundaries. While the domain of application of the model may be defined concretely (i.e. in the sense that we know which entities and relationships can be represented) the model does not similarly hold specifications of what might be explained as a result of its application. Woody further argues that the structure of the model explicitly includes some aspects of *compositionality*. This property relates to the fact that there is a recursive algorithm for the proper application of the model. Hence, while the open boundaries of the model allow its potential application to new, more complex cases, its compositional structure actually provides some instruction for how a more complex case can be treated as a function of simpler cases. Finally, in Woody's framework, a model provides some means of *visual representation*. This characteristic facilitates the recognition of various structural components of a given theory. Many qualitative relations of a theoretical structure can be efficiently communicated in this way. Woody's description of properties of models have implications for science teaching and learning (Erduran, 1999). Table 6.2 provides some examples of learning scaffolds derived from each property, which can act as epistemic criteria in the evaluation of models.

Fostering understanding of the model evaluation criteria communicates that the growth of scientific knowledge is a dynamic process. Students engaging in learning scaffolds such as those illustrated in Table 6.2 would develop appreciation of how models are subject to evaluation, revision and reconceptualization on the basis of a set of criteria. A similar line of reasoning can also apply to various other dimensions of the TLM set. For instance, models and laws can be contrasted with evaluation criteria of consistency and coherence with theories. The key message of the preceding discussion is that enabling students to engage in TLM evaluations can potentially facilitate their understanding of the criteria, standards and heuristics that drive knowledge growth in science.

## 6.5 Explanatory Dimension of Scientific Knowledge

TLM share a special bond: explanation. Explanations are the glue that tie TLM because the intersection of explanatory forms and sources have to jive with one another in addition to standing the rigor of fit with reality (broadly construed).

## 6.5 Explanatory Dimension of Scientific Knowledge

**Table 6.2** Properties of models as evaluation criteria and example learning scaffolds

| Criterion | Verbatim definition (from Woody, 1995) | Learning scaffold |
|---|---|---|
| Approximate | The model is an approximation of a complete theoretical representation for a phenomenon. The model omits many details based on judgments and criteria driving its construction | *What does this model include or exclude about my observations?* <br><br> *Does this model include the theoretical assumptions relevant to these conditions?* |
| Projectable | A model does not come with well-defined or fixed boundaries. While the domain of application of the model may be defined concretely in the sense that we know which entities and relationships can be represented, the model does not similarly hold specifications of what might be explained as a result of its application | *Does this model explain/mimic the phenomenon I am trying to understand?* |
| Compositionality | There is a recursive algorithm for the proper application of the model. Thus, while the open boundaries of the model allows its potential application to new, more complex cases, its compositional structure actually provides some instruction for how a more complex case can be treated as a function of simpler cases | *Can I use this model to understand or explain a new phenomenon?* <br><br> *What features of this model help me understand/explain new phenomenon?* |
| Visual representation | This characteristic facilitates the recognition of various structural components of a given theory. Many qualitative relations of a theoretical structure can be efficiently communicated in this manner | *Does this model have a picture/representation that helps me to see/understand the unknown phenomenon?* |

Explanations are complex entities. They vary in level of complexity, in their ubiquity, and in the way they can be leveraged to illuminate mechanisms underlying phenomena. Explanations vary in scope as well as the purposes for which they are used. They vary in the way they are derived and expressed in specific disciplines (see Dagher & Erduran, 2014, for a discussion on explanations in biology and chemistry).

It is beyond the purpose of this chapter to inquire into all those intriguing aspects of explanation or its coverage in science education (e.g. Braaten & Windschitl 2011; Sandoval & Reiser, 2004). Here the focus is on the notion of "explanatory pluralism" which is ignored in science education despite its utility from scientific and pedagogical standpoints. This specific emphasis is justified on account of two philosophical reasons: (a) Relevance of explanatory pluralism to the concept of family resemblance and interdisciplinarity of science, and (b) relevance of explanatory pluralism to the particular nature of the disciplines and its ability to facilitate communication across disciplinary boundaries. The focus is also justified on account of three pedagogical reasons: (a) Opening spaces for discussion about multiple explanatory frameworks, (b) establishing coherence within different explanatory strands within a given science discipline, and (c) supporting the goal of managing different multi-disciplinary approaches in a problem-based-learning context.

Explanatory pluralism refers to the possibility of having more than one explanation that can explain the same phenomena equally well. It is rooted in a broader conception of epistemological pluralism that "proceeds from empirical observations that the complexity of the natural world eludes complete representation by a single epistemological, theoretical, or investigative approach (Longino, 2002)" (Miller et al., 2008). Explanatory pluralism is typically contrasted with explanatory realism (Grantham, 1999) or explanatory monism where only one explanation is accepted. According to Grantham, these explanations need to meet four criteria:

1. Two distinct explanatory strategies must offer explanations for a single event. Unless these strategies are truly distinct, any case of pluralism will be merely apparent.
2. Both strategies provide correct explanations for the same event. I do not intend to provide a full analysis of what it means to provide a "correct" explanation. Roughly, I mean that the explanation is "true" and complete. Showing that an explanation has heuristic advantages is not sufficient.
3. The explanatory strategies are compatible. To satisfy the realist, both explanations must be true. Thus, it must be possible for particular instances of both strategies to provide true explanations of a single event. Presumably, if the strategies are truly distinct, they will diverge in some cases as well.
4. Neither strategy is "eliminable." That is, neither explanatory strategy can be reduced or completely eliminated from our most complete theory of the domain. (Grantham, 1999, p. S225)

Grantham (1999) demonstrates how two explanatory models can meet the four criteria by applying them to the context of cladogenetic trends. He identifies two distinct styles of explanation that are compatible with each other. One explanation is more active involving focus on "the particular forces that affect each species" while the second explanation involves more "'passive' or 'random' diffusion away from a boundary in morphological space" (p. S223). While the explanations are distinct, trends can be correctly explained in both ways. Further, since neither strategy can be reduced or eliminated, it should be agreed that both strategies can provide correct explanations for a single trend.

While explanatory plurality seems intuitive, there is some uneasiness about it probably emerging from the tacit preference for monistic explanations. This is where understanding acceptable norms in disciplinary practices becomes important from a scientific and pedagogical point of view. Focusing on biology as a case in point, it has been argued that explanations in biology do not aim to provide the typical "necessary and sufficient conditions" as might be expected of a typical explanation in physics. Instead biological explanations aim to "gain partial, but ever increasing insights into the causal workings of various *life processes*" (Brigandt, 2011, p. 262). Mayr's (1961) distinction between proximate and ultimate explanation provides a basic dichotomy between at least two ways of explaining biological systems. In asking about how a phenomenon happens, the proximate explanation would address physiological or other processes that underlie the cause, while the ultimate explanation would address the phenomenon based on the organism's evolutionary history. The explanations do not contradict but rather complement each other by adding a different dimension: one causal and another historical. Thus, compatibility among multiple explanations seems to be more commonly encountered in some disciplines than in others.

Applying this reasoning in other contexts, it is noted that some philosophers address a larger number of possible explanations. For example, Rose (2004) offers a fable that supports the claim that biological systems can sustain a variety of explanations. In this fable, five biologists are having a picnic when they noticed a frog jump into a nearby pond. Posing the question of what caused it to do so, led to five different answers. The physiologist reasoned that impulses traveled from its retina to the brain and then to the leg muscles. The biochemist pointed out the properties of the proteins, actin and myosin, whose fibrous nature enable them to move in a predictable way. The developmental biologist attributed it to the ontogenetic processes that occurred during early stages of cell division. The animal behaviorist attributed the cause to the snake that was lurking by, whereas the evolutionist discussed the role of natural selection in favoring those frogs that escaped their prey due their ability to detect them quickly and move fast in response, allowing them to survive and reproduce. These five different 'orientations' to explanations are valid in their own right. Collectively, they provide a more holistic understanding of the phenomenon under consideration than would be the case if one explanation is provided.

The variety of explanatory frameworks in biology can be captured by a number of explanatory types that support the aim of gaining insights about the "causal workings" of biological systems without limiting their discussion to causal explanations. As discussed in other work (Dagher & Erduran, 2014), Wouters (1995, 2007), for example, outlines six different types of explanation: Physiological, Capacity, Developmental, Viability, and Historical/Evolutionary. These different types of explanation approach the same phenomena from different perspectives. To explain the circulatory system of a given organism, for example, Wouters argues that physiological explanations focus on the types of events in the individual organism's life history, whereas a capacity explanation focuses on underlying causal explanations having to do with the structure of the heart and valves. A developmental explanation would focus on the development of the system (heart and vessels), while a viability explanation would focus on why structural differences between systems occur in different organisms. An evolutionary explanation would focus on differences in systems between organisms in the same lineage. A design explanation, however, is one in which a system in a real organism might be compared to a hypothetical one. Calcott (2009) makes the case for an additional type of explanation that he names, "lineage explanation". This type of explanation aims to make plausible a series of incremental changes that lead to evolutionary change, focusing on a sequence of mechanisms that lead to the successive changes. Lineage explanations "show how small changes between ancestral and derived mechanisms could have produced different behavior, physiology and morphology" (Calcott, p. 74). Consequently, they provide an additional "explanatory pattern" to account for evolutionary change.

The range of explanations described by Wouters (1995, 2007), Calcott (2009), and Rose (2004) illustrates the significance of considering a diverse set of explanations for obtaining a more comprehensive understanding of biological systems. Perhaps one of the overarching attributes of biological explanations is the notion of consilience in which different explanations need not be subsumed under one another and need not contradict with one another. The notion of consilience, attributed to Wilson by Rose, can be viewed as a pragmatic adaptation of the notion of "consilience of inductions" developed by Whewell in his *Novum Organun Renovatum* (Morrison, 2000). The diversity of explanatory types in biology is perhaps reflective of the "epistemological pluralism" (Rose, p. 129) that is characteristic in the study of biological systems.

Appreciation of explanatory pluralism ought to be tempered with the value of explanatory integration (Brigandt, 2008). According to Brigandt, explanatory integration refers to "the integration of ideas and explanations from different disciplines so as to yield an overall explanation of a complex phenomenon" (p. 4). By calling for explanatory integration, Brigandt finds a healthy stance between rampant pluralism and extreme reductionism. From a pedagogical perspective, explanatory integration gives a purpose for explanatory pluralism thus resulting in closer scrutiny of explanatory options and a higher level of discernment in selecting explanatory strands most useful for addressing a given question or solving a particular problem.

Explanatory diversity as well as explanatory integration are often neglected in biology education, not because they are difficult to communicate, but because they are not recognized or valued enough to count as important epistemic companions to cognitive goals. Most school biology textbooks are organized around topics such as cells, tissues, organs and systems. The logical organization is lamented for its lack of attention to the psychological aspects of learning. But textbooks also tend to lack the epistemic aspects of the discipline including the need to create spaces for constructing multiple explanations for the same phenomenon. Furthermore, textbook chapters are focused on biological entities not on key questions that demand reference to multiple explanatory frameworks. Overall, textbooks tend to limit teachers' and students' ability to capitalize on the value of explanatory pluralism as a powerful vehicle for explaining and understanding phenomena in the life sciences or promote explanatory integration for a holistic understanding of how all pieces of the explanation puzzle fit.

The preceding discussion focused on biology has numerous implications for teaching science in general. One could examine disciplinary knowledge for cases of explanatory pluralism and bring this out to students, possibly borrowing the criteria stated by Grantham (1999) as they fit with the target concepts and students' developmental level. Examples such as those just discussed from paleobiology and biology, as well as examples in physics (such as the particle-wave duality issue), chemistry (such as Arrhenius, Bronsted-Lowry and Lewis definitions of acids and alkalis) and other fields could be called upon. A second implication is to distinguish these types of compatible explanations within a particular sub-discipline (i.e. cladogenetics, particle-wave duality), from explanations converging from different sub-disciplines to explain the same phenomena (e.g. jumping frog). The significance of these implications becomes more obvious when considering these ideas in a problem-based-learning environment where students are called to bring to the solution insights from multiple disciplinary perspectives (e.g. math, science, economics etc). Such learning environments would support students in realizing what the explanatory options are and exercising epistemic judgment in selecting those explanations that are most likely to contribute to the solution of the problem under investigation.

## 6.6 Educational Applications

Implications and applications in science education are discussed revisiting the theme of growth of scientific knowledge as illustrated in Fig. 6.1. As discussed earlier, the interrelated TLM cycles into the spiral of scientific knowledge growth, bound by the 'glue' of explanation and leading to understanding. A parallel can be drawn to educational settings. Students can be supported in constructing scientific knowledge in the form of theories, laws and models in a coordinated fashion such that they can appreciate the individual and interrelated components of TLM envisioned in the heuristic as a continually developing coherent and dynamic system. Definitions and "sound bites" of scientific knowledge in terms of particular theories, laws and models are not sufficient for deep understanding of scientific knowledge. How this knowledge comes to be and why specific knowledge claims are supported or not are important for students to learn.

Table 6.3 illustrates some examples of TLM from biology, chemistry and physics in relation to the TLM framework with some example epistemological themes covered in this chapter. As an example, in biology the theory of evolution, models of cells and laws of inheritance work together to enhance understanding of biological traits. The TLM leads to a range of explanations that can be physiological or developmental in nature. The models of the cell or other components like the genes can be visually represented encapsulating the basic components of the genetic mechanisms that underlie the processes of evolution. The TLM framework can increase in complexity across a particular grade level in a sequence of instruction as well as across age levels in different grades. Science curricula can be designed with the coherence of the TLM in mind while at the same time paying attention to "growth of science knowledge" (GSK) issues such as the levels of theories or properties of models. Across the different domains, similar considerations can result in rich and powerful discussions about what makes TLM similar and different across domains of science, thus reinforcing an understanding that science has a 'family' of scientific knowledge forms and at the same time, possesses diversity in its articulation in different disciplines.

In summary, the chapter presented the argument that forms of scientific knowledge need to be communicated to students in a coherent, nuanced and dynamic

**Table 6.3** Growth of scientific knowledge examples in different domains

| Family resemblance | Domain specificity | | |
|---|---|---|---|
| Growth of scientific knowledge issue | Biology | Chemistry | Physics |
| **Levels and kinds** | | | |
| Fringe | Scientific creationism | Alchemy and philosopher's stone | Fifth fundamental force |
| Frontier | Evolution | Nanochemistry | Quarks |
| Center | Cell theory | Periodicity of elements | Quantum theory |
| | Mendelian laws of inheritance | Periodic Law | Planck's Law |
| **Evaluation criteria** | | | |
| Compositionality | Basic features of cells | Bonding and molecular models | Reflection of light rays |
| Visual representation | Pictorial representations of cells | 2 or 3-D models of molecules | Rays diagrams for plane or curved mirrors |
| **Epistemological pluralism** | Developmental explanations | Structural explanations in organic chemistry | Mechanistic explanations |
| | Physiological explanations | Explanations of dynamic reaction mechanisms | Quantum explanations |

fashion to illustrate the nature of growth of scientific knowledge, including key issues such as classifications of scientific knowledge forms, domain-specificity and explanatory pluralism. If learners are expected to have deep understanding about the forms of scientific knowledge, it is imperative that they learn the conceptual aspects in conjunction with the epistemic aspects of scientific knowledge. Understanding the epistemic dimensions of TLM potentially provides students with a richer understanding of why particular scientific knowledge is considered valid, how such knowledge is justified in the first place, and how it can be applied within and across science disciplines.

# References

Aduriz-Bravo, A. (2013). A semantic view of scientific models for science education. *Science & Education, 22*(7), 1593–1611.

Boulter, C., & Gilbert, J. (1996). *Typologies of models for explaining science content*. Paper presented at the annual meeting of National Association for Research in Science Teaching, St. Louis, MO.

# References

Braaten, M., & Windschitl, M. (2011). Working towards a stronger conceptualisation of explanation for science education. *Science Education, 95*(4), 639–669.

Brigandt, I. (2008). *Beyond reduction and pluralism: Toward an epistemology of explanatory integration in biology*. PhilSci Archive. Retrieved from http://philsci-archive.pitt.edu/4329/1/Beyond_reduction_and_pluralism.pdf

Brigandt, I. (2011). Philosophy of biology. In S. French & J. Saasti (Eds.), *The continuum companion to the philosophy of science* (pp. 246–267). London: Continuum Press.

Bruner, J. (1966). *Toward a theory of instruction*. Cambridge, MA: Harvard University Press.

Bunge, M. (1961). Kinds and criteria of scientific laws. *Philosophy of Science, 28*(3), 260–281.

Calcott, B. (2009). Lineage explanations: Explaining how biological mechanisms change. *The British Journal of Philosophy of Science, 60*, 51–78.

Cartwright, N. (1983). *How the laws of physics lie*. Oxford, UK: Clarendon Press.

Chi, M. T. H., Feltovich, P. J., & Glaser, R. (1981). Categorization and representation of physics problems by experts and novices. *Cognitive Science, 5*, 121–152.

Christie, M. (1994). Chemists versus philosophers regarding laws of nature. *Studies in History and Philosophy of Science, 25*, 613–629.

Christie, M., & Christie, J. (2000). "Laws" and "theories" in chemistry do not obey the rules. In N. Bhushan & S. Rosenfeld (Eds.), *Of minds and molecules* (pp. 34–50). Oxford, UK: Oxford University Press.

Christie, M., & Christie, J. (2003). Chemical laws and theories: A response to Vihalemm. *Foundations of Chemistry, 5*, 165–174.

Clement, J. J. (1989). Learning via model construction and criticism. In G. Glover, R. Ronning, & C. Reynolds (Eds.), *Handbook of creativity: Assessment, theory and research* (pp. 341–381). New York: Plenum.

D'Andrade, R. G. (1992). Schemas and motivation. In R. D'Andrade & C. Strauss (Eds.), *Human motives and cultural models* (pp. 23–44). Cambridge, UK: Cambridge University Press.

Dagher, Z., Brickhouse, N., Shipman, H., & Letts, W. (2004). How some college students represent their understanding of the nature of scientific theories. *International Journal of Science Education, 26*(6), 735–755.

Dagher, Z., & Erduran, S. (2014). Laws in biology and chemistry: Philosophical perspectives and educational implications. In M. Matthews (Ed.), *International handbook of history, philosophy and science teaching* (pp. 1203–1233). Dordrecht, The Netherlands: Springer.

Dhar, P. K., & Giuliani, A. (2010). Laws of biology: Why so few? *Systems Synthetic Biology, 4*, 7–13.

Duit, R., & Glynn, S. (1996). Mental modelling. In G. Welford, J. Osborne, & P. Scott (Eds.), *Research in science education in Europe: Current issues and themes* (pp. 166–176). Bristol, PA: Falmer Press.

Duschl, R. (1990). *Restructuring science education: The importance of theories and their development*. New York: Teacher's College Press.

Duschl, R., & Erduran, S. (1996). Modeling the growth of scientific knowledge. In G. Welford, J. Osborne, & P. Scott (Eds.), *Research in science education in Europe: Current issues and themes* (pp. 153–165). London: Falmer Press.

Dutch, S. I. (1982). Notes on the fringe of science. *Journal of Geological Education, 30*, 6–13.

Erduran, S. (1999). *Merging curriculum design with chemical epistemology: A case of learning chemistry through modeling*. Unpublished PhD thesis, Vanderbilt University, Nashville, TN.

Erduran, S. (2007). Breaking the law: Promoting domain-specificity in science education in the context of arguing about the periodic law in chemistry. *Foundations of Chemistry, 9*(3), 247–263.

Erduran, S., & Duschl, R. (2004). Interdisciplinary characterizations of models and the nature of chemical knowledge in the classroom. *Studies in Science Education, 40*, 111–144.

Gable, D., & Bunce, D. (1984). Research on problem solving in chemistry. In D. Gabel (Ed.), *Handbook of research on science teaching and learning* (pp. 301–326). New York: Macmillan.

Geertz, C. (1973). *The interpretation of cultures*. New York: Basic Books.

Giere, R. (1991). *Understanding scientific reasoning* (3rd ed.). Fort Worth, TX: Holt, Rinehart, and Winston.
Giere, R. N. (1999). *Science without laws*. Chicago: University of Chicago Press.
Gilbert, J., & Boulter, C. (1997). Learning science through models and modelling. In B. Frazer & K. Tobin (Eds.), *The international handbook of science education*. Dordrecht, The Netherlands: Kluwer.
Glynn, S., & Duit, R. (1995). Learning science meaningfully: Constructing conceptual models. In S. Glynn & R. Duit (Eds.), *Learning science in the schools: Research reforming practice* (pp. 1–33). Hillsdale, NJ: Lawrence Erlbaum.
Grantham, T. A. (1999). Explanatory pluralism in paleobiology. *Philosophy of Science, 66*, S223–S236.
Grosslight, K., Unger, C., Jay, E., & Smith, C. (1991). Understanding models and their use in science: Conceptions of middle and high school students and experts. *Journal of Research in Science Teaching, 29*, 799–822.
Irzik, G., & Nola, R. (2014). New directions for nature of science research. In M. Matthews (Ed.), *International handbook of research in history, philosophy and science teaching* (pp. 999–1021). Dordrecht, The Netherlands: Springer.
Johnston-Laird, P. N. (1983). *Mental models*. Cambridge, UK: Cambridge University Press.
Justi, R. (2000). Teaching with historical models. In J. K. Gilbert & C. J. Boutler (Eds.), *Developing models in science education* (pp. 209–226). Dordrecht, The Netherlands: Kluwer.
Justi, R., & Gilbert, J. K. (2000). History and philosophy of science through models: Some challenges in the case of 'the atom'. *International Journal of Science Education, 22*(9), 993–1009.
Kuhn, T. (1962/1970). *The structure of scientific revolutions* (2nd ed.). Chicago: University of Chicago Press.
Lakatos, I. (1978). *The methodology of scientific research programmes: Philosophical papers* (Vol. 1). Cambridge, UK: Cambridge University Press.
Lange, M. (2005). Ecological laws: What would they be and why would they matter? *Oikos, 110*(2), 394–403.
Mahner, M., & Bunge, M. (1997). *Foundations of biophilosophy*. Berlin, Germany: Springer.
Mayr, E. (1961). Cause and effect in biology. *Science, 134*, 1501–1505.
McComas, W. (1998). The principal elements of nature of science: Dispelling the myths. In W. McComas (Ed.), *The nature of science in science education* (pp. 53–70). Dordrecht, The Netherlands: Kluwer Academic.
Mihram, G. A. (1972). The modelling process: Transactions on systems. *Man and Cybernetics, SMC-2*(5), 621–629.
Miller, T. R., Baird, T. D., Littlefield, C. M., Kofinas, G., Stuart Chapin III, F. & Redman, C. L. (2008). Epistemological pluralism: reorganizing interdisciplinary research. *Ecology and Society, 13*(2), 46. [online]: http://www.ecologyandsociety.org/vol13/iss2/art46/
Morrison, M. (2000). *Unifying scientific theories: Physical concepts and mathematical structures*. Cambridge, UK: Cambridge University Press.
Press, J. (2009). Physical explanations and biological explanations, empirical laws and a priori laws. *Biology & Philosophy, 24*, 359–374.
Redhead, M. L. G. (1980). Models in physics. *British Journal for the Philosophy of Science, 31*, 145.
Rogers, Y., & Rutherford, A. (1992). Future directions in mental models research. In Y. Rogers, A. Rutherford, & P. Bibby (Eds.), *Models in the mind: Theory, perspective and applications* (pp. 57–71). London: Academic Press.
Rose, S. (2004). The biology of the future and the future of biology. In J. Cornwell (Ed.), *Explanations: Styles of explanation in science* (pp. 125–142). Oxford, UK: Oxford University Press.
Ruphy, S. (2003). Is the world really "dappled"? A response to Cartwright's charge against "cross-wise reduction". *Philosophy of Science, 70*, 57–67.
Sandoval, W. A., & Reiser, B. J. (2004). Explanation-driven inquiry: Integrating conceptual and epistemic scaffolds for scientific inquiry. *Science Education, 88*, 345–372.

# References

Scerri, E. R. (2000). Philosophy of chemistry: A new interdisciplinary field? *Journal of Chemical Education, 77,* 522–526.
Scerri, E. R., & McIntryre, L. (1997). The case for the philosophy of chemistry. *Synthese, 111*(3), 213–232.
Schauble, L., Glaser, R., Duschl, R., Schulze, S., & John, J. (1994). Experimentation in the science classroom. *The Journal of the Learning Sciences, 4,* 131–166.
Shore, B. (1996). *Culture in mind: Cognition, culture and the problem of meaning.* New York: Oxford University Press.
Tversky, B. (1989). Parts, partonomies, and taxonomies. *Developmental Psychology, 25*(6), 983–995.
van Brakel, J. (2000). *The philosophy of chemistry.* Leuven, Belgium: Leuven University Press.
Vihalemm, R. (2003). Natural kinds, explanation and essentialism in chemistry. *Annals of the New York Academy of Sciences, 988*(1), 59–70.
Woody, A. (1995). The explanatory power of our models: A philosophical analysis with some implications for science education. In F. Finley, D. Allchin, D. Rhees, & S. Fifield (Eds.), *Proceedings of the third international history, philosophy, and science teaching conference* (Vol. 2, pp. 1295–1304). Minneapolis, MN: University of Minnesota.
Wouters, A. (1995). Viability explanation. *Biology & Philosophy, 10,* 435–457.
Wouters, A. (2007). Design explanation. *Erkenntnis, 67,* 65–80.

# Chapter 7
# Science as a Social-Institutional System

In this chapter, science is described as a social-institutional system. The system's components include professional activities, scientific ethos, social certification, social values, organizational, political, and financial aspects of science. After reviewing these categories, a framework is introduced that can be extended to educational contexts to guide the teaching and learning of the social-institutional aspects of science. Pedagogical examples are drawn from historical and more contemporary cases in science to highlight some underrepresented features of the social-institutional contexts of science. Some example curricula are used to situate the extent to which the mentioned social-institutional aspects are covered in school science.

## 7.1 Introduction

Science is inherently a social system. It involves individual scientists working in social groups in social institutions, exercising social values and activities. The inclusion of the social dimension of science in science education is warranted for various reasons. First, the ways in which scientists organize science socially might have relevance for how science learning environments can be structured. In other words, students may benefit from acquiring the social aspects of scientific communities, and the inclusion of social features of science in the classroom may facilitate students' learning of science. Second, the particular social values and norms that dominate communities of scientists could be considered as potential learning outcomes for students. To educate students in science does not mean that they will only be acquiring the cognitive and epistemic aspects of science. Understanding science in its entirety will suggest that students learn about the social norms that scientists work by. Without the inclusion of the social context of science in science education, students are bound to have limited understanding of how the scientific enterprise works, and how social structures, relationships and issues influence the development of science. Irzik and Nola (2014) highlight four components of "science as a social-institutional system": professional activities, scientific ethos, social certification and dissemination of scientific knowledge and social values. This chapter builds on theoretical work on these four components and describes additional categories that relate to the political, economic and institutional contexts of science. The institutional aspect of science, though referred to in Irzik and Nola's work, is not specified in detail. In exploring science as a social-institutional system, we will draw on, for instance, some examples from post-colonial accounts of science (e.g. Gupta, 1998) to problematize the particular social and historical contexts of Western science.

Advocating the teaching and learning of science as a social-institutional system is based on the following assumptions. First, raising awareness of various aspects of science, whether they have had a positive or negative impact on society, is important for promoting understanding of science in a way that is consistent with its historical and contemporary practices. This is not to promote the view that the scientific enterprise has been exclusively oppressive but rather that the activities of scientists are not disconnected from societal interests including governmental, political and economic forces. There are some historical and contemporary instances that can be communicated meaningfully as they relate to the topics that are being taught without turning science into social studies. Second, engaging students in social aspects of science (e.g. discussion of colonial oppression through science) promotes ethical awareness and understanding so that oppression and destruction are avoided or at least minimized in future generations. Overall, educators are obligated to present students with a balanced, realistic and holistic account of science that is not overly sterilized nor idealized but rather more reflective of the nature of science as being not only an epistemic and cognitive system but also a social-institutional one.

In the next four subsections, the social categories (i.e. professional activities, scientific ethos, social certification and dissemination, and social values) briefly introduced by Irzik and Nola (2014) are reviewed. Three additional categories (i.e. social organizations and interactions, political power structures and financial systems) are described that are suggested by our review of the research literature. The proposed framework of "science as a social-institutional system" may not necessarily be exclusively limited to these categories but we envisage that it is comprehensive enough to be inclusive of a wide range of issues related to the social dimensions of science such that further articulation by other researchers could be possible in detailing each particular category to emphasize certain dimensions.

## 7.1.1 *Professional Activities*

Scientists do not just produce knowledge but they are engaged in a series of professional activities such as attending conferences, presenting findings, publishing findings, writing research proposals, seeking funding and reviewing papers as well as grant applications (Irzik & Nola, 2014). The professional dimension of scientists' work highlight the fact that scientists are embedded in community practices, aspects of science that are often ignored as learning outcomes in school science. It is often ignored, for instance, that it is through the public sharing and dissemination of results at key professional organizations and peer-reviewed journals that scientists begin to certify and validate their findings. The social and professional context of being a scientist, thus, requires much more than the ability to conduct scientific investigations. Being a scientist requires the skills of professional networking, presenting, writing, financial understanding and critical thinking to evaluate others' work relative to the standards of a community.

Professional activities of scientists contribute to the norms of scientific communities. Modeling such norms in the context of the science classroom is a goal that is consistent with contemporary science education research. Constructivist science teaching methods support the public sharing and presentation of findings in the classroom at the primary/elementary and secondary levels. In fact, it is through such communal sharing and discussion of ideas and appropriate teacher and peer feedback that learners begin to build, shape or revise their ideas in the process of learning. Constructivist science teaching methods also encourage the utilization of argumentation and evidence-based reasoning as foundational elements for developing thinking, arriving at some sense of closure for current investigations (i.e. in a similar manner that scientists would validate their findings with peers) and/or promoting further explorations.

Re-creating professional activities in the classroom, and promoting them as scientific norms, entail drawing explicit parallels as well as distinctions between the students' confined classroom learning community and the more 'open' scientist communities. The process of engaging students in social norms of scientists is useful for honing their skills in 'talking' and 'doing' science, and communicating their findings as well as giving them opportunities to consider viable responses to alternative and contradictory claims raised by other student teams. Contemporary communication technologies provide great opportunities for linking schools in different parts of the world and facilitating the sharing of findings among groups of students working on similar investigations. In the process, online resources, personal blogs, visual images, and data archives can be shared to lend support to students' arguments working within and across communities. Outside the confines of formal education, individuals or teams of students typically participate in similar processes when they take part in science fairs or science Olympiads, where they are asked to justify their findings and explain how they might apply them to novel situations.

In short, there is plenty of opportunities in and out of school for students to participate in activities that are modeled after the professional activities of scientists, raising students' awareness of the ways of acting, thinking and communicating in science.

## 7.1.2 Scientific Ethos

Scientific ethos include the set of institutional "attitudes that scientists are expected to adopt and display in their interactions with their fellow scientists as well as in carrying out their scientific activities" (Irzik & Nola, 2014, pp. 1006–1007). Merton, who derived these norms from interviews with scientists, presented four concepts: universalism, organized skepticism, disinterestedness, and communalism. Universalism affirms that scientific claims get evaluated by being subjected to rational criteria and are not affected by personal factors such as the scientists' nationality, religion or ethnic origin. Organized skepticism involves subjecting claims to scrutiny using scientific reasoning. Disinterestedness refers to scientists' independence from their personal interests and ideologies making it possible for them to reach conclusions that run against their own preferences. Communalism asserts the common ownership of scientific knowledge and openness towards discussion and exchange of ideas and information. These norms are considered as proximate values because they lead to the ultimate goal of arriving at reliable knowledge, and can be considered both descriptive and normative (Allchin, 1999). The Mertonian norms have been criticized by STS scholars due to the increasingly blurred boundaries between public and private science (e.g. Atkinson-Grosjean, 2006).

Resnik's norms (Resnik, 2007 cited by Irzik & Nola, 2014) are considered necessary for the ethical conduct of science. They include the following key concepts: intellectual honesty, respect for research subjects, respect for the environment, freedom and openness. Additional ethical standards discussed by Resnik include integrity, carefulness, openness, respect for intellectual property, confidentiality, responsible publication, responsible mentoring, respect for colleagues, social responsibility, non-discrimination, competence, legality, animal care, and human subjects protection. It is important to state that scientists are expected to adhere to these social norms and that departure from them tends to result in sanctions.

The four Mertonian principles are useful in the context of inquiry-based science teaching. They foster learning of the ideals of subjecting claims to scrutiny; seeking and valuing rational arguments and empirical evidence; and open sharing and validation of ideas. These are both epistemic and social values that need to be experienced and cultivated in school science, and need to be seen as part and parcel of how scientists formulate and evaluate valid claims. This is not to deny that there have been occasional violations of such ethos (such as cases of genuine mistakes, fabricated data, selective and preference bias), but rather it is to affirm that the process of seeking scientific knowledge is not value-free, and that students need to be deliberate about formulating and evaluating scientific claims guided by a set of norms. Scientific ethos should be part of science education because learning science should embody the ethical practices through which this knowledge is constructed. Scientific ethos pertain to exercising good practices in any field of study. Take, for example, intellectual honesty and integrity, respect for intellectual property, respect for the environment, animal care and human subjects protection. Being guided by these ethical codes of conduct allows for knowledge to be trustworthy and to serve the greater good.

### 7.1.3 Social Certification and Dissemination

Scientists engage in investigations that they subsequently put to test through the scrutiny and validation of the scientific community. They collate the results of their investigations and present their work at major conferences and events. They produce manuscripts that they publish in peer-reviewed journals. Through the engagement of the broader scientific community, the work gets reviewed, criticized and evaluated. Irzik and Nola (2014) citing the work of Kitcher (2011), argue that the social system of certification and dissemination of scientific knowledge involves collective and collaborative efforts of the community. The system ensures both a "social quality control" and an "epistemic control" (p. 1008).

Although this account of social certification of scientific knowledge by the broader community incorporates the key mechanisms for certification, it can be considered simplistic. Historically, sharing results of scientific investigations has not always been readily done among scientists. For example, the practices of alchemists were highly secretive and since the scientists of the time were also alchemists, one could argue that some of the scientific practices in history were shrouded in secrecy, and have also avoided the input of the broader scientific community, a condition that violates Merton's norm of communalism. Issac Newton, for instance, observed of Robert Boyle:

> Mr. Boyle has divers [sic] times offered to communicate and correspond with me in these matters but I ever declined it because of his conversing with all sorts of people & being in my opinion too open & too desirous of fame. (quoted in Westfall, 1980, pp. 492–493)

Contemporary practices of scientists are often considered quite secretive as well, and the willingness of scientists to readily share their work is contested (Shapin, 2008). Scientists do try to publish their results first to increase the chance of recognition, and are constantly in competition to ensure that they are not beaten to the dissemination of their results (Marshall, 2002). They manage the dissemination of information strategically to ensure credit through publication (Latour & Woolgar, 1979). Less traditional publication venues such as electronic journals have been accompanied by innovative sharing of data banks and scientific methods. For example, the *Journal of Visualized Experiments* (available online at http://www.jove.com) conveys how scientific methods in some instances are being certified visually, illustrating the effective use of emerging communication technologies in reporting and certifying scientific investigations.

Engaging students in the social certification and dissemination processes of science is crucial in facilitating students' understanding that scientists historically and currently engage in communication practices, which at times might be prone to secrecy and competition. It is important to distinguish between the normative and the empirical claims to openness and secrecy. In principle and ideally, there should be dissemination of science and no secrecy. In practice, however, this norm is sometimes violated. It is likely that such issues will emerge naturally at the level of the classroom given the human inclination about possessiveness of intellectual work. Teachers could capitalize on similar patterns of behavior exhibited by students to illustrate both the positive and negative aspects of social certification and dissemination processes, for instance via sharing of group work and more widely across classrooms or even schools, in sharing science projects. Computer technologies could be used to mediate the sharing, validation and evaluation of student produced projects.

### 7.1.4 Social Values of Science

Respecting the environment, social utility and freedom are considered social values that are embodied by science. Freedom is necessary for pursuing scientific research, a value that is validated by historical accounts of Western Science (e.g. Jacob, 1997). Increased attention to the social utility of scientific research is necessary for garnering public support. Respect for the environment is critical for human survival. Freedom and respecting the environment as social values serve also as ethical principles as noted in the sub-section on scientific ethos.

Students, who as future citizens are expected to be called upon to participate in making public decisions in and about science, should be aware that scientific pursuits can be stifled by imposed ideological or religious constraints. They need to recognize the necessity of a cultural climate of freedom for promoting scientific research, appreciate the contributions of basic and applied research to the public good, and ensure that in the process of generating or applying scientific knowledge no harm is done to the environment.

## 7.2 Elaborating on Science as a Social-Institutional System

There are additional components of science as a social system that go beyond the ideas explored in the previous section. These include perspectives from, for instance, cultural studies of science. "Cultural studies of science" is a multidisciplinary research field drawing from sociology, anthropology, feminist theory, and history and philosophy of science. A key precept in this field is that scientific practices are not only situated in cultural, historical and social contexts but also that understanding science requires an interpretive and critical engagement with science. Within the science education research community, the cultural studies of

## 7.2 Elaborating on Science as a Social-Institutional System

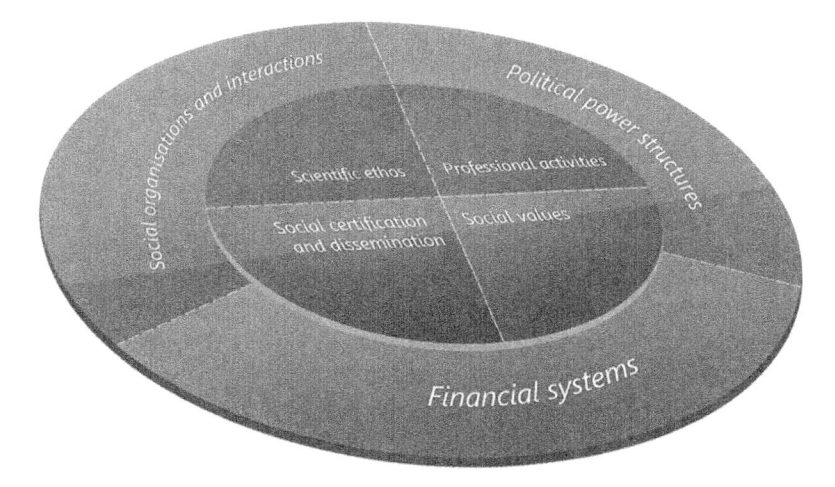

**Fig. 7.1** Science as a social-institutional system

science have been interpreted and applied by numerous colleagues (e.g. Bencze, Sperling, & Carter, 2012; Roth & Middleton, 2006; Scantlebury, 2008; Tobin & Roth, 2007). Some journals have dedicated whole special issues to the explication of how, for example, worldviews impact science (e.g. *Science & Education* in 2009). There is an entire journal (i.e. *Cultural Studies of Science Education*) dedicated to research on cultural studies in science education.

Given the vast body of work in the area of cultural studies of science in science education, the aim here is not to replicate what is already reviewed elsewhere. Instead, we focus on three broad categories that extend Irzik and Nola's (2014) notion of science as a social and institutional system, and consider their pedagogical significance for achieving a broader understanding of science in context. These categories are "social organizations and interactions", "political power structures", and "financial systems". In summarizing the main features of science as a social-institutional system then, a visual representation is proposed (see Fig. 7.1) which summarizes the main features that are explored in this chapter. In each sub-section, some example implications for science education are presented to illustrate relevance for science teaching and learning.

Categories of science as a social-institutional system can be visualized in terms of (a) the core features of professional activities, scientific ethos, social certification and dissemination and social values, and (b) the broader features of political power structures, financial systems and social organizations and interactions. The latter features are referred to as broad because finance, politics and institutions are integral components of the larger society in which science, like other organized human activity, is being practised. In reality, however, all categories of this system are interactive with porous boundaries. Furthermore the number of categories can be increased (and therefore the size of each 'slice' be adjusted) to consider another relevant social-institutional aspect not represented here.

In the context of science teaching and learning, the time allocated to each category need not be the same for every topic or issue discussed. Rather such details would be determined by the curricular relevance and student interest. When teaching a science topic, the teacher could consider questions such as: To what extent can collaboration in science be addressed in this lesson/unit? What scientific norms need to be emphasized? Does this topic lend itself to including credibility of scientists? Who benefits from scientific knowledge? Who is harmed by it?

Engaging students in professional activities similar to what scientists do allows them to get a flavor of authentic practices that can support their understanding of the social context of science, though in a limited way. Typical activities include working in teams, and presenting and debating findings. However, such activities rarely include attending meetings outside the parameters of the classroom, publishing findings, reviewing research papers, 'grant' applications, writing research proposals and seeking funding. Of course our reference to such aspects of the scientific enterprise often unusual in school science does not mean that students would be engaged in them in a professional sense. Rather, for instance, student grant applications can be based on major undertakings involving industry, business or government funders. There are available opportunities for high school students in the USA to apply to federally sponsored research by the National Science Foundation, 9–12 Program funded by the US-Army research Labs, or STEM grants sponsored by private businesses. If such opportunities are not available or participation in them is not possible, alternatives can be developed locally whereby modest small school budgets could be set up to help model for students how financial considerations can impact the design of investigations.

There have been some initiatives in which communication technologies are used to share data across project sites and promote the kind of team spirit that are seen in networks of scientists. A classic example is the National Geographic Kids Network which has involved students from around the world. Schools from all 50 states in the USA and 52 countries are said to have participated in these units (see http://www.nationalgeographic.com/kidsnetwork/index.html). However, such practices remain limited in most schools due to the high level of coordination needed across school sites and the costs involved. Teachers may not be very familiar with the strategies that would enable the coordination of those social-institutional aspects of science as learning aims at the level of science lessons. However, considering how science investigations that cannot be done in the classroom (and where resorting to second hand investigations becomes the next best thing) a similar principle can be applied in relation to some social-institutional aspects of science as well. For instance, once or twice a year the teacher can present students with second hand reports about a current or historical debate amongst scientists that is relevant to their field of study and explore with them how it was resolved using a select number of original science papers and media reports. Discussing both types of writing genres would potentially enable students to develop stronger level of functional literacy and to be exposed to the formal genre of scientific reports as well as the popular versions of those accounts.

Having described each of the four categories of science as a social-institutional system outlined by Irzik and Nola (2014), we now turn to articulate additional categories that are deemed to be important to capture in school science. The categories are "social organizations and interactions", "political power structures" and "financial systems".

## 7.2.1 Social Organisations and Interactions

Scientists work in institutions like universities, research centers and industrial sites that are socially organized. Within each institution, teams of scientists work on particular projects. For example, in her analysis of the professional and employment status of CERN researchers, Knorr-Cetina (1999) illustrates the following classification of researchers: student, postdoc (employed by outside institutes), fellow (employed by CERN), outside staff (employed by outside institutes) and CERN staff (employed by CERN). She then explains that such classification gets people sorted relative to their career stage (student, postdoc, senior person), their employment status (staff or non-staff), and their source of funding (whether or not he or she is financed by CERN). There is thus an organizational hierarchy that then dictates the nature of the social interactions among the members of the team. Knorr-Cetina provides an empirical account on, for example, how the different members of the team establish trust:

> Trust distinctions do not form a single taxonomy; rather, they underlie and are mixed into several classifications. First, they are used informally to sort people into those "whose results one can believe" and who one wishes to cooperate with, and those one does not. Second, they are superimposed upon formal classifications designating the physicists' professional status. And third, they draw the important distinction between expert and non-experts. (Knorr-Cetina, 1999, p. 131)

She provides excerpts from CERN physicists on how they reflect on their interactions with their colleagues. Trust "functions as a sort of selection mechanism that brings some individuals together and keeps them connected in confidence pathways" (Knorr-Cetina, 1999, p. 202). The pathways in turn help form links across individuals relative to groups, institutes and experiments.

Beyond the level of interactions among scientists working in a particular social institution, one can also consider the wider organizational contexts of science. The scientific enterprise has been closely related to the military (Kaiser, 2002) and the industry (Kleinman, 1998). Academic science itself is increasingly connected to business interests through funding sources. Many scientists themselves have established firms and encouraged collaboration of businesses with academics (Powell, 1996). The study of commercialization of life sciences, for example, illustrate the dichotomy of collaborative research relationships versus market transactions (Powell, 1990). The broader institutional contexts of science thus create further levels of organizational complexity that situates science as an enterprise and the interactions of scientists with a wide range of stakeholders including the defense sector, the industry and the academy.

Such analyses provide a nuanced way of understanding how scientists work within and across social organizations and how they interact with each other as well as with stakeholders. The implication for science education is that such literature on the institutional organizations and cultures of science can inform understanding of science as an enterprise. Reference to the institutional dimension of science raises awareness of the complexity of team memberships, work, division of labor and social activities that surround the lives of scientists. Some of the work in developmental psychological accounts of learning related to community practices, including the socio-cultural notions of peripheral participation and zone of proximal development (e.g. Lave & Wenger, 1991; Wenger, 1998) are applicable to scientific settings, and thus have direct implication for structuring learning contexts in schools. In other words, the role of peers and distributed expertise are not only important from a cognitive point of view, but they already are operational within organizational structures and interactions of the scientific enterprise itself as illustrated by philosophical and institutional accounts of science. In short, the category of "social organizations and interactions" provides a perspective on what it means to be a scientist as an employee or an employer, and how the institutional structures, dynamics and politics shape and form the interactions among scientists working in and across organizations.

### 7.2.2 Political Power Structures

The coverage of science in science education tends to focus on the benefits and universality of scientific knowledge and the progress science has contributed to the human condition. This has led to the construction of a narrative of science as a value-free endeavor, a knowledge-base that enables countless advancements in new technologies that can counteract disease or support the exploration of distant planets. Technology, and not science, is typically blamed for negative outcomes whether it is the pollution resulting from dumping chemical waste in the environment or the fatalities resulting from the deployment of atomic bombs.

Such a sterile account of science, traditionally deemed appropriate for school science, is problematic for a number of reasons. It creates an artificial boundary between (a) scientific knowledge and the scientists who produce it, (b) science and technology, and (c) science and the governments and states that support it. This misrepresentation of the context of science consequently leads to downplaying the 'cozy' relationship between science and state patronage. Both science and technology have been historically linked to governments and states, advancing, for instance, their colonial interests. Consider Galileo sharpening his telescope to better identify distant enemy fleets (Fermi & Bernardini, 2003) and Heisenberg's contribution to Hitler's scientific projects (Rose, 2002) serving as tools for oppression, intimidation, or justification for intervention.

The interplay of science with politics of governments, race as well as gender among other factors has now been extensively studied. Feminist philosophers of science such as Sandra Harding have argued for the consideration of the role of gender in the characterization of science (e.g. Harding & Hintikka, 1983), a perspective that has also received criticism due to claims being made about the contributions of feminist epistemology to science (Pinnick, 2005). Feminist philosophers such as Evelyn Fox Keller argue against the "ideology of gender" that has served to exclude women from the world of science (1996). This ideology of gender works in obscure ways at different stages of development, and usually conveys or reflects societal biases that influence parental decisions (e.g. what counts as appropriate books and toys for their children), and teacher and parent expectations (e.g. children's abilities and aptitudes). Along with gender politics in application to analyses of science, a particular body of scholarship focuses on how science served as a vehicle for colonial domination. Londa Schiebinger defines "colonial science" as

> …any science done during the colonial era that involved Europeans working in a colonial context. This includes science done in Europe that drew on colonial resources in addition to science done in areas that were part of Europe's trading or territorial empires. (Schiebinger, 2005, p. 52)

The ideological power of science was spread to the colonies through the various European powers and indeed Western scientific knowledge was co-constituted with colonialism (McLeod, 2000). Numerous branches of science have been studied in their colonial context. Studies on ecology (Anker, 2001), astronomy (Pang, 2002) and medicine (McLeod, 2000), for example, provide insight into how science both served colonial ambitions and was also itself constituted through colonialism.

Particular examples illustrate how science has been used by colonial forces. For example, the issue of cholera in India and how it was situated within the British colonial discourse provides one case study. There were specific connections between medical theories and conflicting agendas for the best ways of maintaining political control over the colonies. The identification of the causes of cholera was directly relevant to the British ambitions of control and governance in India. Cholera theory and the question of locality played a role in justifying the governance of India (Arnold, 1986, 1993). If the factors contributing to cholera were pathogens, then they could be studied by scientists outside of India. If, however, they were environmental, then this would give the British authority to establish precedence for governing India. Another example involves botany and the role of visual culture during the Spanish Enlightenment. Images played a significant role in the Spanish colonial ambitions. In the hands of the Spanish colonial powers, botanical imagery not only served as instruments for natural investigations but also advanced rationales for oppression of local cultures. Images of indigenous 'commodities' like people, plants and animals were transported from their origin of composition and production, and thus became global commodities, justifying the Spanish presence in faraway lands for the so-called scientific ideals (Bleichmar, 2012).

Apart from the use of science as an ideological tool by European forces in colonizing and the co-constitution of science through colonialism, science was also disseminated to the natives for particular purposes, namely in positioning the natives as inferior and less powerful, thus maintaining the power imbalance between the colonizer and the colonized. Through science exhibits, for example, the British demonstrated to the natives how much they lacked in their knowledge. Exhibits were set up to legitimize British power and project justice through public dissemination of newly discovered scientific knowledge (Prakash, 1999).

Science education has the responsibility to unveil how scientific knowledge can become a tool for oppression and exploitation to countless victims when co-opted to serve gender, colonial, economic or other interests, and in the process alienate individuals or groups like women, de-humanize communities, destroy ecologies and cultures. Lack of acknowledgement of these connections makes it possible for future citizens, scientists included, to repeat past transgressions as they are caught unaware of the ramifications of their actions. Science education should thus uncover the political heritage of science and move beyond a naive conceptualization of science that perpetuates a legacy of injustice.

Understanding how science relates to inter and extra-state politics, as shocking as it may seem, is important for helping students develop a critical sense of scientific literacy without undermining the importance, value, or benefits of scientific knowledge and rationality. Students need to understand these relationships in ways that allow them to view science in context and to develop a more sophisticated understanding of how scientific knowledge is generated, used, and at times abused. Students should understand how history of science includes political ends that go beyond the ideal goal of constructing explanatory synthesis about the natural world.

## 7.2.3 Financial Systems

The actions of scientists and the distribution of resources in science are mediated by economic forces. States and governments around the world have governing bodies such as the National Science Foundation in the USA and the Science Foundation Ireland that provide funding for research to universities and research institutions where science is done, thereby influencing the nature of scientific research conducted. In order to carry out investigations, scientists need resources for which they need to bid funding, operating within the standards and expectations of funding agencies as well as the scientific community:

> Insofar as research funding agencies become important actors in these systems of monitoring and ranking research groups and institutes, they obviously gain additional authority over scientists. As the key organisations in the competitive award of research funds, they often become the central state supported bodies responsible for both influencing priorities towards public policy goals and judging the merits of proposals and scientists bidding for funding. However, their strategic autonomy and capabilities depend on the level of state delegation of resources and administrative procedures to their officials, on the one hand, and their dependence on scientific elites in making judgments, on the other hand. (Whitley, 2011, p. 372)

There is now a body of literature (e.g. Wibble, 1998) that analyzes science from the perspective of its financial and economic context given that "science has ceased to be considered as a system of available knowledge outside the economic circuit" (Salomon, 1985, p. 79). Recent arguments for situating science in its financial context follow earlier observations that "…the community of scientists is organized in a way which resembles certain features of a body politic and works according to economic principles similar to those by which the production of material goods is regulated" (Polanyi, 2002/1969, p. 465).

Commodification and commercialization of science are significant issues that have been attracting increasing attention by the science education community although very few studies have taken economics of science as an explicit theme relative to science education research (e.g. Erduran & Mugaloglu, 2013). Radder (2010) defined commodification as "the pursuit of profit by academic institutions through selling the expertise of their researchers and the results of their inquiries" (p. 4). The commercial nature of science is related to (a) the production of scientific knowledge as private property (as opposed to the public ideal), and (b) the existence and development of science as a market which hinders free consumption of scientific knowledge by its public and/or rival producers.

One observation in economics of science concerns the relationship between science and technology. Scientific knowledge is produced by using scientific methodology, an approach that has gained credibility and status because it is purported to offer a reliable means to knowledge (see Chap. 5 and Chap. 6). The produced scientific knowledge is then considered a valuable commodity that can be put on sale, sometimes as a technological output and sometimes in itself as a scientific innovation such as a particular procedure to produce genetically modified foods. The close impact of science on technology is one of the driving forces of economics of science, given that technology is itself grounded in notions of productivity, growth, commodities and markets (Diamond, 2008). Technology and innovation are conceived to contribute to new profit opportunities by creating demand for new products, potentially also decreasing the cost of production. The role of the scientist in this scenario is one of a producer or supplier of scientific information. A scientist is also an employee with a salary derived from a university, a research institution (non-profit or for-profit) or industry. In some instances, individual scientists can establish their own spin-off companies to patent and mass-produce a particular method or product.

The inclusion of economics perspectives in science education ensures that learners of science are equipped with the skills to understand that science has a financial dimension. The goal of financial intelligence in science education serves at least two goals. First, for those students who are going to go into science career routes, it provides them with the awareness that science is not an insular body of knowledge but rather that there is an institutionalized system that is tied to economic factors and political agencies. The chances of students' success in performing to the expectations of the research funding culture of academic and research institutions will thus be facilitated early on in their enculturation into science. Second, for those students who may not choose to have science careers but rather aim to serve as informed and educated public citizens, it is crucial to recognize how their own contributions to state economy are being used to fund scientific research.

The journal *Science & Education* dedicated a special issue to the commercialization and commodification of science (Irzik, 2013). Some papers in the special issue illustrate how the literature on economics of science can be applied in science education. For example, Erduran and Mugaloglu (2013) discuss how the financial interpretation of the traditional invention versus discovery dichotomy can be portrayed in science teaching and learning through argumentation. The existing curricular content can potentially provide numerous other examples for generating resources. For instance, while studying the DNA structure and advances in genomics, the link to gene technology and patent rights lends itself to deliberations on how science and economics interact, including the landmark decision by the United States Supreme Court to rule against the patenting of the human gene (Liptak, 2013).

## 7.3 Educational Applications

So far, we have developed Irzik and Nola's (2014) characterization of the social-institutional system of science with reference to professional activities, scientific ethos, social values, and social certification and dissemination, and extended this framework to include the features of social organizations and interactions, political power structures and financial systems (Fig. 7.1). The definition of each category has been elaborated, and their relevance for science education has been highlighted. Even though the framework generated is not exhaustive and can be further expanded with more categories, the emergent framework provides some guidelines for pedagogical use. In this section, the implications for science education are discussed in more detail. In order to render the earlier reviews meaningful and useful for educational purposes, a concept map is generated (see Fig. 7.2) summarizing some of the key categories of science as a social-institutional system.

### 7.3.1 Teaching and Learning of Science as a Social-Institutional System

How might science be taught and learned as a social-institutional system in the classroom? Can we, as science educators, avoid turning science lessons into social studies in so doing? Such questions have often confronted colleagues who have advocated the incorporation of a societal dimension in science teaching such Science-Technology-Society (STS) (e.g. Aikenhead, 1994), socioscientific issues (e.g. Zeidler, Sadler, Simmons, & Howes, 2005) or re-contextualization of science learning to promote civic engagement and citizenship (e.g. Bencze et al., 2012). All these movements have highlighted the significance of the inclusion of the social contexts of science in science education. Despite some of the nuanced differences in perspectives among some of the proponents of these movements, we believe that our approach is, broadly speaking, in line with the key agenda which is that science teaching and learning have to situate science in its social context, and have to demonstrate that the social dimension is just as integral to science as the cognitive and epistemic dimensions. In considering the question posed earlier regarding how science lessons would differ from social studies lessons when they incorporate the social contexts of science, our view is that a balance is needed across

7.3 Educational Applications 151

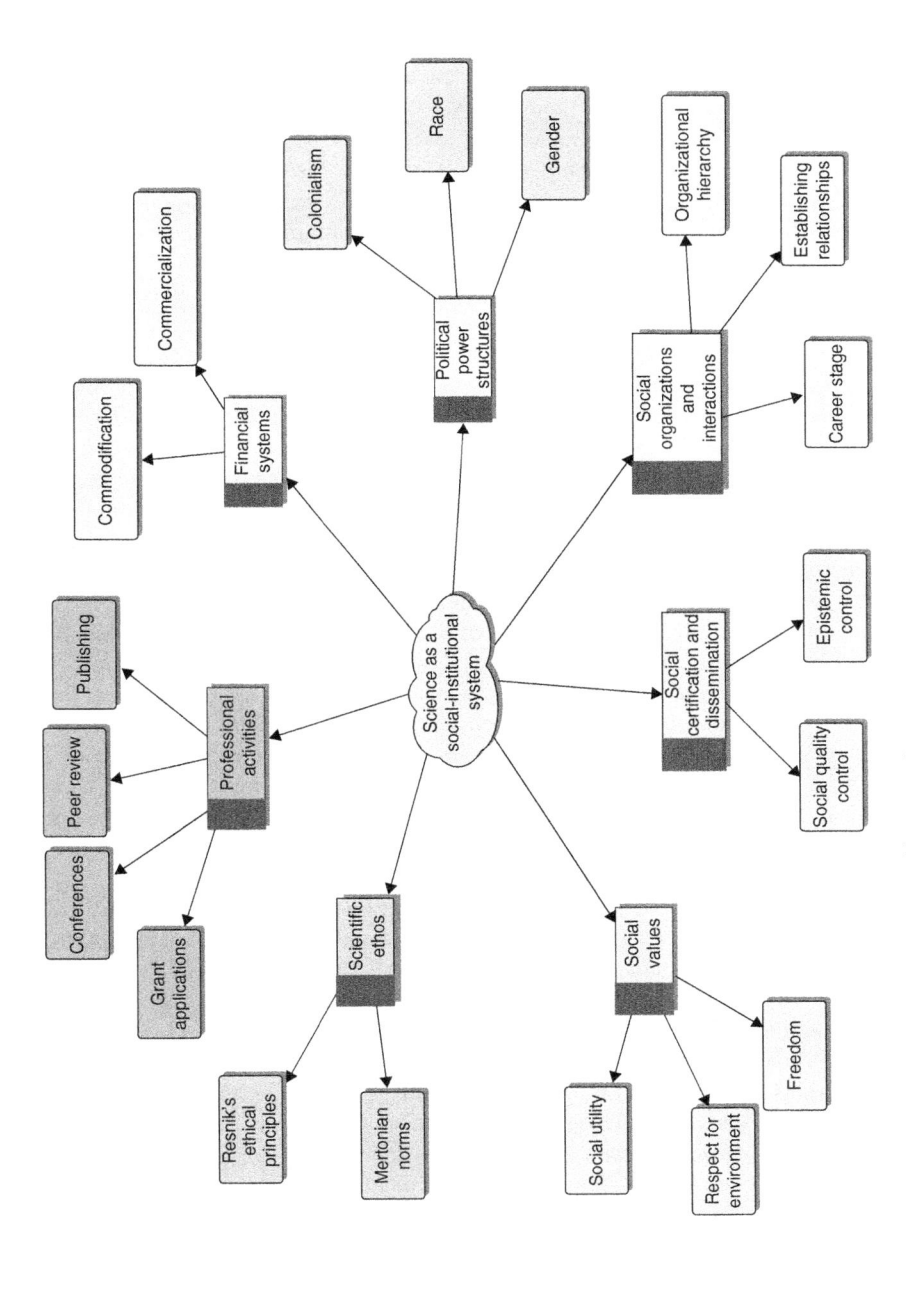

**Fig. 7.2** Science as a social-institutional system and its categories

the various aspects of science as we have been highlighting in each chapter of this book. The inclusion of science as a social-institutional system is not to be done at the expense of science as a cognitive and epistemic system. In this sense, a problem does not arise with crowding an already crowded curriculum with more goals. Rather, a coordinated effort is needed to ensure that the existing curriculum is balanced and configured so as to touch on the various aspects of science. It may be that not all aspects are covered in one single lesson or a single unit. As discussed in Chap. 8, the teaching and learning of science in a holistic fashion does not necessarily require so much the addition of new content but it requires better articulation and balance of NOS goals and careful selection of curriculum materials that support the attainment of these goals.

Take for example, the topic of genetic engineering and the related issues surrounding genetically modified organisms. This topic is a standard component of most high school biology curricula in many parts of the world. The *scientific and technological relevance* of genetic engineering for solving medical problems (e.g. targeting cancer cells) and improving crops are often discussed in biology curricula to illustrate the importance of these concepts in attending to everyday problems. The *economic/civic* aspect regarding funding of academic and industrial research in this area raises questions about the outcomes of the research, who owns the knowledge, and who owns patents to inventions. Even though the economic/civic aspect leads to insights into current events that give rise to new legislation about what is patentable and what is not, it is often discussed the least in science lessons. In addition, the *values* that govern and enter into the selection of research subjects (e.g. humans and animals) provide another aspect that is worthy of consideration but are rarely discussed. Understanding only the scientific and engineering aspects of how genetic engineering is done will not be sufficient to prepare students in understanding the multifaceted dimensions of the topic. The societal, economic, environmental and value dimensions related to genetic engineering are not frills but are closely intertwined with scientific research. Components of science as a social-institutional system are essential in understanding science and thus should be made more visible in the science curriculum.

Let us now turn to some examples to illustrate collectively the various categories of the social context of science (i.e. science as a social-institutional system, sometimes referred to in this chapter as "social context" for brevity). The examples are intended to bring our theoretical discussion closer to some practical classroom level implementation. The concept map introduced in Fig. 7.2 can be used to organize some theoretical ideas in a visual representation. The concept map has the potential for use as a pedagogical tool. In other words, by distilling the key ideas from the theoretical discussion, an organizing framework can be generated that can have some intellectual utility for science education researchers as well as for teachers and students to organize their thinking. From a pedagogical perspective, the concept map can (a) be used as a planning device to help teachers organize social aspects of science that they cover in lessons; (b) be used as an instructional resource to help students categorize key concepts and unpack each concept as well as the relationships between these concepts within a unit of study or across units.

One historical and one contemporary case from science are used to illustrate how a subset of concepts can be highlighted by building the cross-links noted in Fig. 7.3. In each case, how particular concepts are interlinked and what educational implications they offer are presented. The case is briefly outlined, and discussed. The examples illustrate how each scenario can be transformed for pedagogical purposes including implications for learning. Subsequently, connections across the various categories of science as a social-institutional system are highlighted on the concept maps.

### 7.3.1.1 Case 1 – Linnaeus' Classification of Organisms and its Relation to the Whaling Industry

The example is based on Linnaeus' classification of organisms particularly fish and whales. It illustrates a number of ideas pertaining to the relationship between science, financial systems and political structures (see Fig. 7.3). In March 1818, new laws were passed in the United States to ensure the quality and proper taxation of fish oils, and a team of inspectors was charged with the task of collecting fines on un-inspected fish oil. Four months later, when an inspector, James Maurice, tried to collect fines on three uninspected caskets of fish oil imported by James Judd, the latter admitted not paying the tax because he believed that whale oil was not subject to the same rules governing fish oil, asserting that whales were different from fish. A court case, Maurice *v* Judd, followed in which Judd's claim that whale oil is not fish oil was contested. Expert witnesses, including the naturalist and physician Samuel Mitchell, testified that whales do not belong in the same category of fish. After a 3 day trial and a 15-min jury deliberation, the case was closed in favor of Judd. A month later, the law was amended to exclude from inspection whale-related products. This case ultimately sided with Linnaeus' classification system that was counter-intuitive at the time from a lay-person's perspective. The court case challenging classification of whales as mammals shows how scientific knowledge can sometimes be contested and used to serve state and commercial interests. Whereas the outcome was justified from a scientific standpoint, it certainly helped that the scientific facts of the case were also allied with the financial interests of the merchants of whaling goods (Burnett, 2007). It turns out that this issue of classification (see Chap. 4 for a more extended discussion of classification as scientific practice) is not trivial as it gets at the heart of a key scientific practice that is not set in stone, but constitutes an organizational process based on certain assumptions about nature. It is believed that Linnaeus himself originally thought that whales were fish, and later decided they were mammals.

There are many lessons to be learned from this story. How scientists classify organisms can be epistemically grounded in some assumptions. These assumptions, as logical as they may seem, may not also square well with common-sense ways of viewing the world. Scientific systems of classification can impact legislation policies and have economic implications (for another example, see the tomato court case dating back to 1893, http://caselaw.lp.findlaw.com/cgi-bin/getcase.pl court=us&vol=149&invol=304). This simple example addresses some important ideas that are accessible to secondary and tertiary students.

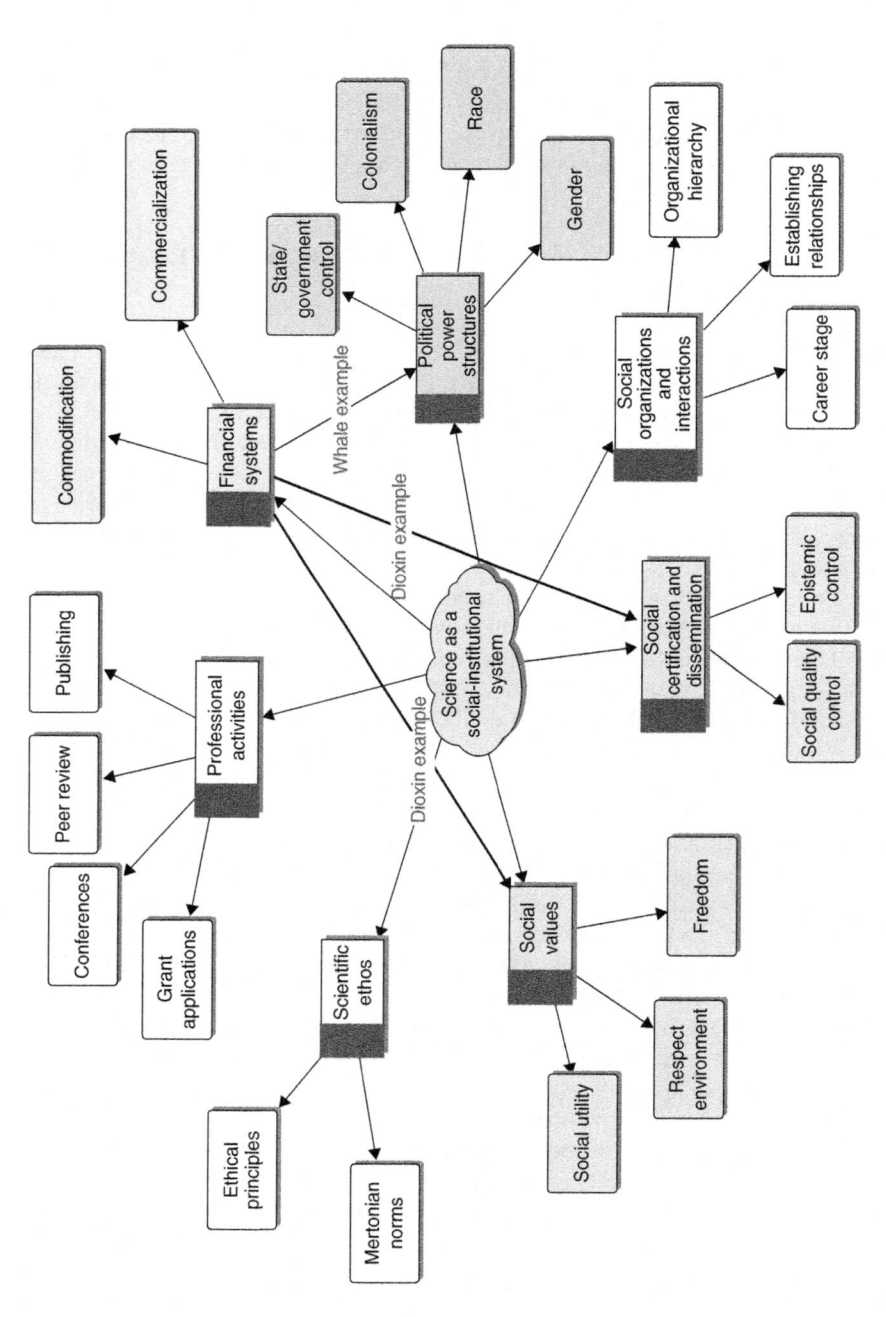

**Fig. 7.3** Linnaeus' classification and epidemiology of dioxin as examples in "science as a social-institutional system"

7.3 Educational Applications

### 7.3.1.2 Case 2 – Epidemiology of Dioxin

The example highlights the concepts of "financial systems" and "social values" (Fig. 7.3). In particular, the case of epidemiology of dioxin as discussed by Machamer and Douglas (1999) is reviewed. The case engages with the issues related to profit making in industry, social interactions between scientists and their employers as well as the values of social utility and respect for human life. Dioxins are chemicals that are by-products of many industrial processes and they are very toxic even in small doses. The danger of dioxins to human health is not contested although the amount of how much is toxic has been debated for many years. Experiments with animals in the laboratory have indicated that dioxins can lead to birth defects, cancer, and reduced immunological response. In the United States, the National Institute for Occupational Safety and Health initiated a dioxin registry to trace workers who were exposed to dioxin-contaminated herbicides (Fingerhut et al., 1991). The resulting study had a large sample size and rigorous methodology, representing a comprehensive account of the epidemiology of dioxin.

The significance of the results of Fingerhut et al.'s (1991) study were soon disputed by Collins, Acquavella, and Friedlander (1992) on the basis that Fingerhut et al. study's conclusion about Soft Tissue Sarcomas (STS) was not reliable. Machamer and Douglas (1999) question the motivations of Collins and colleagues by highlighting that all three individuals worked for a chemical company responsible for dioxin pollution. Machamer and Douglas state: "One might expect that the company's interest in profit-making would determine which outcomes are acceptable in their employees' work" (Machamer & Douglas, p. 49). The social values of protection of the employer and profit does not center in the debates. Neither do conflicts of interest between the needs of the company and the needs of public health, which demand a closer look.

The two examples lend themselves to pedagogical exploration using the basic structure of the concept map as a heuristic to navigate the connections. For instance, in the context of studying taxonomy, students could be presented with the case study scenarios of Linnaeus' classification of whales and the epidemiology of dioxin to identify which aspects of science as a social-institutional system are embedded in these stories. Through identification of these aspects in relation to the stories as well as comparison of the different links across concepts, understanding of the various categories can be reinforced. Students could be engaged in discussions about current scientific events to help them identify how scientific information corresponds to the concept map. For example, when discussing news about new drugs, students can go beyond the headlines and find out who sponsored the research (e.g. drug companies vs. academic), who benefits from it (e.g. patients and companies), the extent to which the findings are certified or still under trial (e.g. patents and patent disputes). They could be encouraged to (a) identify cases where they clarify a target set of concepts that they are interested in, (b) depict instances that illustrate the various aspects of science as a social-institutional system, (c) co-develop a concept map based on their research and extend it by adding new categories, sub-categories, or cross-links, and (d) find examples to illustrate those social-institutional dimensions of science. The use of the entire concept map with each example is likely to promote understanding of science as a social-institutional system. The concept 'state/ government' is derived from the historical story and was not part of the original concept map presented in Fig. 7.3 which was derived from the literature review. This aspect of our discussion highlights the potential of the map (and indeed the broader theoretical framework on "science as a social-institutional system") to be generative and inclusive of related concepts.

## 7.3.2 Curricular Policy

To what extent do curriculum policy documents address the social-institutional dimensions of science? How does the level of detail addressed in this chapter compare to the level of detail included in these documents? By detail, we mean those key concepts that outline science as a social-institutional system. For example, we can ask if a set of curriculum standards make reference to the idea of the political power structures that are embedded in science. To answer these questions we consulted three science education curriculum standards documents, two from the USA spanning 17 years (in order to capture historical differences), and one from England. The purpose for this review is to identify the level of specificity pertaining to science as a social-institutional system. The *National Science Education Standards* [NSES] (NRC, 1996) published in the USA outlines eight science content standards. Three of the eight standards pertain to science disciplines: physical science, life science, earth and space science, while the other five standards are about science and include: Unifying concepts and processes (includes concepts that cut across science), science as inquiry, science and technology (includes design), science in personal and social perspectives, and history and nature of science. Focusing on the history and nature of science standard, the document focuses on three main ideas or sub-standards: science as a human endeavor (perhaps addressing ideas that would hit under science as a social system), nature of scientific knowledge, and historical perspectives.

More recently in the USA, the *Next Generation Science Standards* [NGSS] (NGSS Lead States, 2013) are based on a vision set in *Framework for K-12 Science Education* (NRC, 2012). Unlike NSES, the NGSS focuses around three dimensions that are tightly interwoven throughout the document: Scientific and engineering practices, cross-cutting concepts and scientific ideas, whereby each standards include these three dimensions. The NGSS incorporates practices and cross-cutting concepts in each of the content standards, thus making them part and parcel of teaching the science content. The NGSS considers nature of science to be integrated into the scientific practices and cross-cutting concepts. However it also details some of the big ideas about nature of science (NOS) in Appendix H, where it spells out eight major NOS categories and outlines how four of them relate to scientific and engineering practices, and how the other four relate to cross-cutting concepts. Of the latter set, "science as a human endeavor" is one category that is most relevant to the contents of this chapter. However several references are spread throughout the document itself.

At the time of the writing of this book, a new science curriculum was being developed in England. The draft of the new curriculum has been available. Unlike the 2007 curriculum, the February 2013 version of *Science: Programme of Study for Key Stage* 4 (DfE, 2013) does not seem to play a strong emphasis on the "How Science Works" component which encapsulated some of the categories in Table 7.1 primarily under the heading of "Applications and Implications of Science". Indeed, there has been a vast amount of public dissatisfaction in the formulation of the 2014 national curriculum in England as the content is deemed to revert to an emphasis on subject knowledge (e.g. http://www.theguardian.com/commentisfree/2013/feb/12/round-table-draft-national-curriculum). Furthermore, it should be noted that DfE (2013) is more analogous to the NRC's documents and are general guidelines as educational targets. They do not have the level of specificity captured in documents such as NGSS. Such level of detail is provided in schemes of work that are generated by "exam boards" such as AQA and EdEdcel which provide the instructional as well as assessment resources based on the national curriculum.

By comparing how the three documents address the social aspects of science, it is observed that some of the ideas discussed in our framework on science as a social-institutional system are clearly addressed in the standards and some are not. Focusing on the recent NGSS document, it is evident that fewer ideas about science as a social-institutional system are addressed in Appendix H (see Table 7.1). There is no reference to professional activities, scientific ethos, social certification and dissemination, social organization and interactions. Also there is vague reference to the idea that science and society influence one another without specific reference to political structures and science as a financial system. It is possible to address these influences of science on society and society on science without touching on political and financial structures. In contrast, note how QCA (2007), a previous curriculum document in England, refers explicitly to social, economic, and environmental effects. (The question marks in Table 7.1 indicate that we could not discern an explicit reference for a given category in the documents.)

While the NGSS document does not explicitly address some of the categories discussed in this chapter, the parent document *Framework for K-12 Science Education* (NRC, 2012) includes more elaborate reference to scientific norms, collaborative nature of scientific work, and the larger social and economic factors as seen in the following excerpt:

> Finally, science is fundamentally a social enterprise and scientific knowledge advances through collaboration and in the context of a social system with well-developed norms. Individual scientists may do much of their work independently or they may collaborate closely with colleagues. Thus, new ideas can be the product of one mind or many working together. However, the theories, models, instruments, and methods for collecting and displaying data, as well as the norms for building arguments from evidence, are developed collectively in a vast network of scientists working together over extended periods. As they carry out their research, scientists talk frequently with their colleagues, both formally and informally. They exchange emails, engage in discussions at conferences, share research

**Table 7.1** Components of science as a social-institutional system in example science curriculum standards from the USA and England

| Categories of science as a social-institutional system | USA<br>NSES, 1996 (NRC, 1996)<br>*History and nature of science* [Standard 8] | USA<br>NGSS, 2013 (NGSS Lead States, 2013)<br>*Science as a human endeavor* [Category-Appendix H] | England<br>(QCA, 2007)<br>*How science works* |
|---|---|---|---|
| Professional activities | "Scientists value peer review" (p. 200) | ? | ? |
| Scientific ethos | "Scientists have ethical traditions" (NRC, 1996, p. 200)<br><br>… "Violations of such norms do occur, but scientists responsible for such violations are censured by their peers.". (p. 201) | ? | Pupils should be taught "…about the use of contemporary scientific and technological developments and their benefits, drawbacks and risks" (p. 223) |
| Social certification and dissemination | "Scientists value peer review, truthful reporting about the methods and outcomes of investigations, and making public the results of work". (pp. 200–201) | ? | Pupils should be taught "…how uncertainties in scientific knowledge and scientific ideas change over time and about the role of the scientific community in validating these changes" (p. 223) |
| Social values | "Scientists are influenced by societal, cultural, and personal beliefs and ways of viewing the world. Science is not separate from society but rather science is a part of society." (p. 201) | "Individuals and teams from many nations and cultures have contributed to science and to advances in engineering."<br><br>"Scientists' backgrounds, theoretical commitments, and fields of endeavor influence the nature of their findings.". (p. 6) | "…pupils learn about the way science and scientists work within society" (p. 221) |

## 7.3 Educational Applications

| Social organization and interactions | ? | ? | ? |
|---|---|---|---|
| Political power structures | ? | "Science and engineering are influenced by society and society is influenced by science and engineering." (p. 6) | ? |
| Financial systems | ? | "Science and engineering are influenced by society and society is influenced by science and engineering." (p. 6) | Pupils should be taught "…to consider how and why decisions about science and technology are made, including those that raise ethical issues, and about the social, economic and environmental effects of such decisions" (p. 223) |

techniques and analytical procedures, and present and respond to ideas via publication in journals and books. In short, scientists constitute a community whose members work together to build a body of evidence and devise and test theories. In addition, this community and its culture exist in the larger social and economic context of their place and time and are influenced by events, needs, and norms from outside science, as well as by the interests and desires of scientists. (NRC, 2012, p. 27)

While the excerpt acknowledges the social context of science, it seems to emphasize the social interactions among the scientists. In other words, the social context is mainly about social interactions. The alternative view that is advocated in this book, considers the social context of science to be an inherent part of science that might be manifested not only in the social interactions of scientists but also at higher levels of societal organization like political and economic systems. The social context needs to be unpacked relative to the various social-institutional dimensions of science for clarity of the relevant factors involved in particular cases. It is important for curriculum developers and teachers to seek out opportunities to include the social-institutional dimensions of science as part of a more authentic and comprehensive account of science for science teaching and learning.

## 7.4 Conclusions

In this chapter, we have extensively elaborated on Irzik and Nola's (2014) characterization of science as a social-institutional system in terms of professional activities, social certification and dissemination, scientific ethos and social values. We then extended such characterization to be more broadly inclusive of the political, organizational and financial aspects of science by proposing three additional categories which we labeled as "social organizations and interactions", "political power structures" and "financial systems". The whole set of categories are represented in a visual tool that can guide the production of instructional and learning resources for science education purposes. The visual representation was used to generate concept maps, which can serve in turn as teaching and learning tools. The investigation into science education policy in the USA, and England suggest that some social-institutional aspects of science are underrepresented.

The emphasis on the social and institutional aspects of science provide novel opportunities to redefine and characterize the teaching and learning of NOS, particularly at an age of globalization of knowledge economies that demand more sophisticated understanding of nature of science relative to its various dimensions. Science can be made more authentic in school science by emphasizing that scientists work in communities of practice and that their interactions are governed by particular social norms, values and forces. The inclusion of the sociological, political, organizational and financial contexts of science in science education is likely to engage to students from diverse backgrounds, and improve their interest and engagement in science.

# References

Aikenhead, G. S. (1994). Consequences to learning science through STS: A research perspective. In J. Solomon & G. Aikenhead (Eds.), *STS education: International perspectives on reform* (pp. 169–186). New York: Teachers College Press.

Allchin, D. (1999). Values in science: An educational perspective. *Science & Education, 8*, 1–12.

Anker, P. (2001). *Imperial ecology: Environmental order in the British Empire, 1895–1945*. Cambridge, MA: Harvard University Press.

Arnold, D. (1986). Cholera and colonialism in British India. *Past and Present, 113*, 118–151.

Arnold, D. (1993). *Colonizing the Cody: State medicine and epidemic disease in nineteenth-century India*. Berkeley, CA: University of California Press.

Atkinson-Grosjean, J. (2006). *Public science, private interests: Culture and commerce in Canada's networks of Centres of Excellence*. Toronto, ON, Canada: University of Toronto Press.

Bencze, L., Sperling, E., & Carter, L. (2012). Students' research-informed socioscientific activism: Re/Visions for a sustainable future. *Research in Science Education, 42*(1), 129–148.

Bleichmar, D. (2012). *Visible empire: Botanical expeditions and visual culture in the Hispanic Enlightenment*. Chicago: University of Chicago Press.

Burnett, G. H. (2007). *Trying Leviathan: The nineteenth-century New York court case that put the whale on trial and challenged the order of nature*. Princeton, NJ: Princeton University Press.

Collins, J., Acquavella, J. F., & Friedlander, B. (1992). Reconciling old and new findings on dioxin. *Epidemiology, 3*(1), 65–69.

Department for Education. (2013, February). *Science: Programme of study for key stage 4*. Retrieved August 4, 2013, from http://media.education.gov.uk/assets/files/pdf/s/science%20-%20key%20stage%204%2004-02-13.pdf

Diamond, A. M. (2008). Economics of science. In S. N. Durlauf & L. E. Blume (Eds.), *The new palgrave dictionary of economics* (2nd ed., pp. 328–334). Basingstoke, Hampshire: Palgrave.

Erduran, S., & Mugaloglu, E. (2013). Interactions of economics of science in science education and implications for science teaching and learning. *Science & Education, 22*(10), 2405–2425.

Fermi, L., & Bernardini, G. (2003). *Galileo and the scientific revolution*. Mineola, NY: Dover.

Fingerhut, M., Halperin, W., Marlow, D. A., Piacitelli, L., Honchar, P., Sweeney, M., et al. (1991). Cancer mortality in workers exposed to 2,3,7,8 Terrachlorodibenzo-p-Dioxin. *New England Journal of Medicine, 324*(4), 212–218.

Fox Keller, E. (1996). *Reflections on gender and science*. New Haven, CT: Yale University Press.

Gupta, A. (1998). *Postcolonial developments: Agriculture in the making of modern India*. Durham, NC: Duke University Press.

Harding, S. G., & Hintikka, M. (Eds.). (1983). *Discovering reality: Feminist perspectives on epistemology, metaphysics, methodology, and philosophy of science*. Dordrecht, The Netherlands: Reidel.

Irzik, G. (2013). Introduction: Commercialization of academic science and a new agenda for science education. *Science & Education, 22*(10), 2375–2384.

Irzik, G., & Nola, R. (2014). New directions for nature of science research. In M. Matthews (Ed.), *International handbook of research in history, philosophy and science teaching* (pp. 999–1021). Dordrecht, The Netherlands: Springer.

Jacob, M. (1997). *Scientific culture and the making of the industrial west*. New York: Oxford University.

Kaiser, D. (2002). Cold War requisitions, scientific manpower, and the production of American physicists after World War II. *Historical Studies in the Physical and Biological Sciences, 33*, 131–159.

Kitcher, P. (2011). *Science in a democratic society*. New York: Prometheus Books.

Kleinman, D. L. (1998). Pervasive influence: Intellectual property, industrial history, and university science. *Science and Public Policy, 25*(2), 95–102.

Knorr-Cetina, K. (1999). *Epistemic cultures: How the sciences make knowledge*. Cambridge, MA: Harvard University Press.

Latour, B., & Woolgar, S. (1979). *Laboratory life: The social construction of scientific facts*. Beverly Hills, CA: Sage.

Lave, J., & Wenger, E. (1991). *Situated learning: Legitimate peripheral participation*. Cambridge, UK: Cambridge University Press.

Liptak, A. (2013, June 13). Justices, 9–0, bar patenting human genes. *The New York Times.* Retrieved June 13, 2013, from http://www.nytimes.com/2013/06/14/us/supreme-court-rules-human-genes-may-not-be-patented.html?_r=0

Machamer, P., & Douglas, H. (1999). Cognitive and social values. *Science & Education, 8,* 45–54.

Marshall, E. (2002). Data sharing. DNA sequencer protests being scooped with his own data. *Science, 295*(5558), 1206–1207.

McLeod, R. (Ed.). (2000). *Nature and empire: Science and the colonial enterprise* (Osiris, Vol. 15, pp. 1–13). Chicago: University of Chicago Press.

National Research Council. (1996). *National science education standards.* Washington, DC: National Academies Press.

National Research Council. (2012). *A framework for k-12 science education.* Washington, DC: National Academies Press.

NGSS Lead States. (2013). *Next generation science standards: For states by states.* Appendix H. Retrieved from http://www.nextgenscience.org/next-generation-science-standards

Pang, S. K. (2002). *Empire and the sun: Victorian solar eclipse expeditions.* Stanford, CA: Stanford University Press.

Pinnick, C. L. (2005). The failed feminist challenge to fundamental epistemology. *Science & Education, 14,* 103–116.

Polanyi, M. (2002/1969). The republic of science: Its political and economic theory. From knowing and being. Reproduced in P. Mirowsky & E. M. Sent (Eds.), *Science bought and sold: Essays in the economics of science* (pp. 465–485). Chicago: University of Chicago Press.

Powell, W. W. (1990). Neither market nor hierarchy: Network forms of organization. *Research in Organizational Behavior, 12,* 295–336.

Powell, W. W. (1996). Interorganizational collaboration in the biotechnology industry. *Journal of Institutional and Theoretical Economics, 120,* 197–215.

Prakash, G. (1999). *Another reason: Science and the imagination of modern India.* Princeton, NJ: Princeton University Press.

QCA. (2007). Science: The programme of study for key stage three and attainment targets. *Science: National Curriculum*, London.

Radder, H. (Ed.). (2010). *The commodification of academic research: Analyses, assessment, alternatives.* Pittsburgh, PA: University of Pittsburgh Press.

Resnik, D. (2007). *The price of truth.* New York: Oxford University Press.

Rose, P. L. (2002). *Heisenberg and the Nazi atomic bomb project, 1939–1945: A study in German culture.* Berkeley, CA: University of California Press.

Roth, W.-M., & Middleton, D. (2006). The making of asymmetries of knowing, identity, and accountability in the sequential organization of graph interpretation. *Cultural Studies of Science Education, 1,* 11–81.

Salomon, J. (1985). Science as a commodity-policy changes, issues and threats. In M. Gibbons & B. Wittrock (Eds.), *Science as a commodity* (pp. 78–98). Harlow, UK: Longman.

Scantlebury, K. (2008). Whose knowledge? Whose curriculum? *Cultural Studies of Science Education, 3*(3), 694–696.

Schiebinger, L. (2005). Forum introduction: The European colonial science complex. *Isis, 96,* 52–55.

Shapin, S. (2008). *The scientific life: A moral history of a late modern vocation.* Chicago: University of Chicago Press.

Tobin, J., & Roth, W.-M. (2007). *The culture of science education: Its history in person.* Rotterdam, The Netherlands: Sense Publishers.

Wenger, E. (1998). *Communities of practice: Learning, meaning, and identity.* Cambridge, UK: Cambridge University Press.

Westfall, R. S. (1980). *Never at rest: A biography of Isaac Newton.* Cambridge, UK: Cambridge University Press.

Whitley, R. (2011). Changing governance and authority relations in the public sciences. *Minerva, 49,* 359–385.

Wibble, J. R. (1998). *The economics of science: Methodology and epistemology as if economics really mattered.* London: Routledge.

Zeidler, D., Sadler, T., Simmons, M., & Howes, E. V. (2005). Beyond STS: A research-based framework on socioscientific issues education. *Science Education, 89,* 357–377.

# Chapter 8
# Towards Generative Images of Science in Science Education

In this chapter, the contributions of the Family Resemblance Approach (FRA) to reconceptualizing the nature of science for science education are visited collectively having been detailed individually relative to particular categories in previous chapters. The following questions are raised: How are the various FRA categories related to curriculum standards? How can science learning be supported in developing understanding of holistic accounts of NOS? It is argued that the FRA categories bring coherence to the content of NOS in the science curriculum when coupled with effective teaching strategies. Having proposed in previous chapters specific visual tools to ease memory, conceptualization and communication of the FRA categories, we now refer to them collectively as the *Generative Images of Science* (GIS) to emphasize their pedagogical utility. The FRA and the GIS heuristics are applied to example curriculum standards. In concluding the book, further contributions of the FRA to research and development in science education are explored and some recommendations are offered.

## 8.1 Introduction

So far in this book, the Family Resemblance Approach (FRA) to the characterization of science has been expanded in order to illustrate its potential for applications in science education. By so doing, it has been argued that a new perspective on NOS can provide a platform for developing a holistic and a more inclusive model of science for science teaching and learning. A particular feature of the approach has been the formulation of visual tools that can represent various aspects of science. Visualization is stressed due to its potential to create tangible conceptual representations for relatively abstract concepts. The significance of visualization in science teaching and learning has been extensively reported in science education research literature (e.g. Gilbert, 2005; Gilbert, Reiner, & Nakhleh, 2008;

Johnson-Laird, 1998; Phillips, Norris, & Macnab, 2010; Wu, Krajcik, & Soloway, 2001). In Chap. 2, we have reconfigured the FRA to be represented as a wheel that can be memorable as a comprehensive representation of the various features of science, including the categories that we have generated to supplement those of Irzik and Nola (2014), which were elaborated on in Chap. 7. In Chap. 3, we summarized a simple triangle distinguishing the epistemic, cognitive, and social aims and values of science. In Chap. 4, we proposed the Benzene Ring heuristic to highlight the dynamic nature of the epistemic, cognitive and social components of scientific inquiry. Similarly Chap. 5 offered the 'gears' image to illustrate how explanatory consilience is achieved through the coordination of evidence obtained from different methods. The growth of scientific knowledge framework in Chap. 6 provided yet another form of visual representation of the dynamic nature of the growth of scientific knowledge and its forms as theories, laws and models. Chapter 7 presented a pie chart to represent the social and institutional categories of science. The variety of social contexts was displayed in terms of pieces of a pie, each of which claim space in the science curriculum depending on its relevance to the science content covered at a given time. In Fig. 8.1 we bring together the different representations created in each chapter following a theoretical review, which provides the foundation of their coherence, content, and justification. Collectively, we refer now to these images as "Generative Images of Science" (GIS) since each of them has the potential to be extended and embellished, yet have some central aspects of science captured relative to each component of the FRA.

The images are 'generative' due to their potential to be unpacked and extended for further articulation both from philosophical and pedagogical points of view. Some of the extensions can include ideas that would illustrate aspects of science in a generic sense while others might be more domain-specific. In either case, the FRA enables the articulation of the issues in the sense of a 'family' category. A balancing act is struck between the domain-general and domain-specific aspects of science. In each chapter of the book, we have illustrated how the particular GIS relates to the science curriculum and how each might be potentially used in instruction. Beyond the theoretical articulation, the adaptations of GIS can act as heuristics for teacher

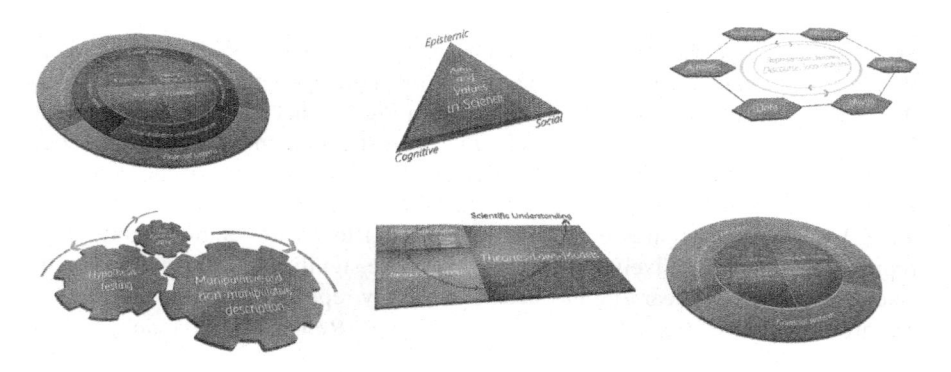

**Fig. 8.1** Generative Images of Science (GIS)

8.1 Introduction

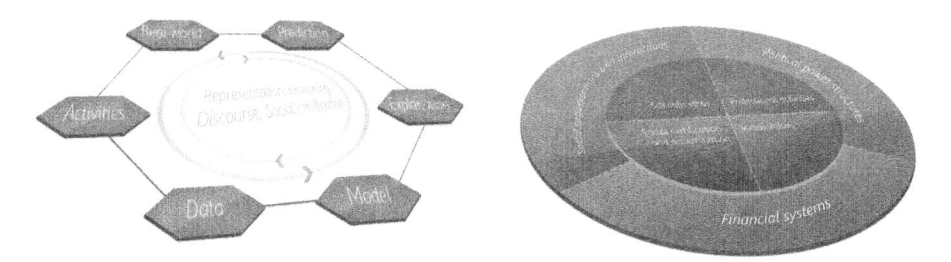

**Fig. 8.2** Potential interactions between GIS: scientific practices and social certification

educators, teachers as well as students in capturing a particular aspect of science (e.g. scientific knowledge) as well as science in its overall comprehensive depiction as illustrated with the FRA wheel in Chap. 2. For example, pedagogical adaptations of GIS could potentially act as meta-cognitive tools in communicating to teachers by teacher educators and to learners by teachers the various components of science. GIS could also potentially form a coherent and comprehensive account to formulate assessment criteria such that new assessments can be developed for investigating NOS understanding.

The GIS which are theoretically grounded and justified, are envisaged as interactive components of the FRA model of NOS. In being interactive and dynamic, they possess the potential to generate and highlight new links between them. For example, as illustrated in Fig. 8.2, the Benzene Ring heuristic of scientific practices in Chap. 4 and the pie chart of the social and institutional aspects of science captured in Chap. 7 can be interlinked. The Benzene Ring heuristic illustrates the epistemic, cognitive and social dimensions of scientific practices as being intricately linked in one holistic representation. The links between the different epistemic components are established by the dynamic socio-cognitive processes represented by the electron cloud denoting representation, reasoning, discourse and social certification. The internal ring structure represents the 'cloud' of social processes (including the sociological, cultural and economic dimensions) that anchor the epistemic components. The links between the different GIS can also be made more explicit. For instance, some components, of the social-institutional system discussed in Chap. 7, for instance the idea of "social certification", can be directly imported into the articulation of the ways in which scientific practices like modeling operate, for instance through peer review.

In the rest of this chapter, the discussion is grounded in two current curriculum policy documents. The recently published USA-based *Next Generation Science Standards* (NGSS) document is used to illustrate how the FRA relates to these curriculum standards. The discussion articulates areas that match with the new standards and others where no match was found. In this case, supplementary coverage is proposed. The choice of NGSS is justified for two reasons: (1) they have been recently published and in that sense capture current thinking about science education priorities in the USA, and, (2) earlier science education reform efforts in the

USA have tended to influence much of curriculum reform efforts around the world. Reference is also made to the *Science: Programme of Study for Key Stage 4* (DfE, 2013) from England to illustrate how FRA ideas work with another curriculum policy that is structured differently. The readers could find the FRA analysis of sample curriculum policy documents useful enough to motivate them to apply a similar analysis to documents that are of immediate relevance to them. So our purpose is not to provide an exhaustive overview of how the FRA apply to curricula internationally but rather to illustrate how the FRA can inform the analysis of science curriculum goals.

## 8.2 Educational Applications of FRA and GIS

How can the FRA and GIS be used in educational contexts? The question is addressed through a series of illustrations. As a reminder, the 'wheel' with the various categories of science as a cognitive-epistemic and social-institutional system presented in Chap. 2 is the basic image on which the other images are developed. In other words, aims and values of science, scientific practices, methods and methodological rules, knowledge, and social-institutional systems are all embedded within the wheel. Hence Fig. 2.1 is the primary tool on which the instructional applications are based with the potential to unpack the various categories through the other GIS (see Fig. 8.1).

From a curriculum planning perspective, the main task while translating the contents of the 'wheel' to practice is to maintain attention on *all of its different components* when planning units of study. The ideas on the wheel can be addressed at the elementary, middle, and high school levels because they involve reflective thinking on science concepts. Through the wheel, students can ask questions that connect what they are already doing, the methods they are using, and the knowledge they are producing. The time allocated to attending to each category depends on its relevance to the context and content of the grade level and unit. Just as the complexity of science concepts unfolds as students move from primary to secondary schooling, so does the complexity of the ideas about science that can be culled to enrich the learning of these concepts along the FRA categories. In other words, it is possible to select strands of ideas from all of the dimensions of the FRA, as detailed in the previous chapters, in a relevant and developmentally appropriate manner to students of all ages. As long as they are made relevant to target science concepts, there exist, by necessity, multiple strands of ideas in each of the FRA categories that can be brought to bear on the topic (Fig. 8.3).

In connecting elements of the wheel to focus on target science concepts, we reaffirm that we are not advocating a particular curriculum approach. Using a basic science, Socio-Scientific Issues, Science-Technology-Society, history of science or any other framework to guide curriculum development, it is appropriate to apply the components of the wheel that fit in best with the content. The design constraints are determined in part by the science content focus and contextual relevance for students. Both of these areas – the content and the design constraints – work together

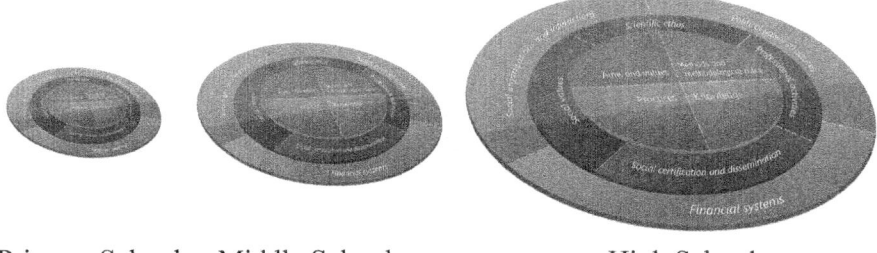

Primary School    Middle School                High School

**Fig. 8.3** The FRA categories get a larger share of coverage as science concepts increase in complexity across grade levels

to aide the selection of the components that could be emphasized in different parts of the curriculum.

To illustrate the way these theoretical ideas translate into practice, the horizontal and vertical articulation of the FRA components in the science curriculum are considered. As is commonly known among curriculum designers, the idea of "vertical articulation of scope and sequence sets its analytical sight on cross-grade concerns. It is the tool used to build coherence in the educational experience of children during their entire school career" (Kridel, 2010, p. 771). In contrast, in the context of horizontal articulation "...scope and sequence has to do with how school experiences early in academic career will logically and coherently flow into experiences offered later in the year" (Kridel, p. 771).

## 8.2.1 Vertical Articulation

In a conventional science curriculum, science concepts are articulated vertically by ensuring that basic exposure to these concepts is implemented early in the primary grades and is developed as students progress from kindergarten to high school. This progression can be noted in many curriculum guides and can be followed in Table 8.1 in relation to the topic of "Heredity: Inheritance and Variation of Traits" obtained from the NGSS (NGSS Lead States, 2013a) as an example. Here we see how basic understandings about this topic are developed across the years along a developmental pathway where a deeper understanding is targeted at the high school level.

Since the focus of the discussion is on considering the FRA ideas relative to the science curriculum, we describe how this can be done at the primary and the secondary levels in relation to the specific content of heredity and variation as an example. The process detailed in the following paragraphs can provide some suggestions for curriculum developers on how to engage with nature of science (NOS) based on a FRA model with the same content across stages of schooling.

For analytical purposes, the FRA categories are listed in the same order that was discussed in previous chapters in the book. This should not be interpreted to mean that

**Table 8.1** Standards on heredity: inheritance and variation of traits, based on the NGSS (NGSS Lead States, 2013a)

| K-2 | Elementary school (G. 3–5) | Middle school (G. 6–8) | High school (G. 9–12) |
|---|---|---|---|
| Heredity: inheritance and variation of traits | Inheritance and variation of traits: Life cycles and traits | Growth, development, and reproduction of organisms | Inheritance and variation of traits |
| 1-LS3-1. Make observations to construct an evidence-based account that young plants and animals are like, but not exactly like, their parents | 3-LS3-1. Analyze and interpret data to provide evidence that plants and animals have traits inherited from parents and that variation of these traits exists in a group of similar organisms | MS-LS3-1. Develop and use a model to describe why structural changes to genes (mutations) located on chromosomes may affect proteins and may result in harmful, beneficial, or neutral effects to the structure and function of the organism | HS-LS3-1. Ask questions to clarify relationships about the role of DNA and chromosomes in coding the instructions for characteristic traits passed from parents to offspring |
|  |  | MS-LS3-2. Develop and use a model to describe why asexual reproduction results in offspring with identical genetic information and sexual reproduction results in offspring with genetic variation | HS-LS3-2. Make and defend a claim based on evidence that inheritable genetic variations may result from: (1) new genetic combinations through meiosis, (2) viable errors occurring during replication, and/or (3) mutations caused by environmental factors |
|  |  |  | HS-LS3-3. Apply concepts of statistics and probability to explain the variation and distribution of expressed traits in a population |

the ideas should be addressed in the same sequence, but rather that the purpose here is to facilitate systematic comparison across primary and secondary science. Ideally, the questions listed under the FRA dimensions are embedded in investigations.

### 8.2.1.1 Primary Science Example

Taking the topic of variation and diversity, young children as early as kindergarten explore such questions as: "What are all the living things we can find in a small plot? Are they the same? Are they different?" (Chalfour & Worth, 2006). At this young age, students are typically guided through a number of investigations that include looking for organisms, observing their characteristics, drawing them, then lumping them into

8.2 Educational Applications of FRA and GIS

broad groups (plants, animals), further examining each group and organizing them into broad categories such as eight legged ones are spiders, six legged ones are insects. In the process of performing observations and discussing findings, the teacher can use the "FRA wheel" (Fig. 2.1) to select which aspects of each of the main FRA dimensions are best to emphasize. Those are then prioritized and embedded strategically in the course of lessons. The main task is to select components from each dimension that best fit in with target science concepts and in ways that enhance student engagement. The following outline provides an example starting point in designing lessons that would cover each of the FRA categories at the primary level.

Aims and values: Focus on accuracy, critical examination, revising convictions, and not harming organisms or destroying plans in the process of observing them (i.e. respect for the environment).
Practices: Focus on reflecting on types of representation, differences in data collection and organization and their affordances, considering the multiplicity of patterns, developing models for grouping organisms, focusing on features of classification models.
Methodology: Focus on elements of observations, what things to look for, the idea that students are engaged in observations and need not manipulate the observed entity. As children engage in the activity they are asked to reflect on the different ways in which they observed it (only their eyes, magnifying lens). They consider how did the use of tools (or not) affect their observation. They speculate on how might a scientist (e.g. botanist, entomologist) observe a similar kind of terrain? What tools would she use to make sense of the findings?
Knowledge: Focus on the structure of the knowledge produced. The teacher would go beyond the answers to the initial questions to ask the students about what was learned, why would this information help them do and how? Why might scientists care to do similar investigations? What happens to the knowledge they come across and how do they use it? How do they coordinate the information they use about biodiversity to arrive at theories?
Social-institutional system: Focus on scientific ethos, professional activities, social certification in relation to ecological issues. How does what students did in class resemble what scientists do? How do scientists establish their findings when examining the same question? Do they change their ideas? Have they always classified things the way the students did? Here things like pharmaceuticals and ecological diversity may come up, even contributing to film making (e.g. http://bugsaremybusiness.com/bio.htm), which children would find fascinating. Issues of biotic diversity and how they affect decisions on land use can be brought in such as building a bridge, or a shopping center and how it might impinge on species diversity (e.g. local cases can be linked to this if deemed appropriate).

#### 8.2.1.2 High School Science Example

Using the topic of genetic variation and diversity, high school students can explore such questions as: How does genetic diversity affect the persistence or decline of a species? What causes genetic variability? How does this variability affect the survival of a species? What solutions can address issues of endangered species? An inquiry into these questions can begin by focusing on the context of humans and later explore them in terms of populations and ecosystems. Students can be guided through a number of investigations that include making observations about what characteristics they share with family members and then with classmates, looking for shared and unique characteristics within family, then discuss their observation about their classmates, and people in the larger society. Students are encouraged to use observations about traits or phenotypes to make claims about genotypes, and explore through multiple simulations and actual data how variation in parent genotypes results in variation in phenotypic and genetic variation in offspring. Eventually this continues on to relating genes to chromosomes, study of genomics and so on. In the process of participating in a myriad of investigations, the teacher can use the FRA wheel (see Fig. 2.1) to select which aspects of each of the FRA dimensions are best to emphasize. As is the case with the previous example, the teacher will find several big NOS ideas that they can choose to emphasize as they go through the unit. The following is an example outline for specifying each category of FRA applied to the topic.

Aims and values: Focus on accuracy of observations, critical examination, revising convictions, and not harming the organisms in the process of observing them (respect for the environment). These aims and values are similar to those covered in earlier grades but can be expanded upon in greater complexity and sophistication.

Practices: Focus on reflecting on their observations, data collection and organization, finding patterns, reflecting on statistical models they for predicting offspring, focusing on the relevance of patterns and anomalies they note in the application of the models.

Methodology: Focus on elements of observations, note the contributions of observations to explanations that are not observable. As they investigate, students reflect on the different ways in which observations in this domain can be done (with and without tools). They consider how scientists use manipulative tools to study the human genome. They compare the methods they used to those used by scientists.

Knowledge: Here the focus could be on the structure of the knowledge produced. Going beyond the scientific knowledge, students explore how the gene concept evolved over time and understand the role of models and theories in shaping knowledge growth. This would lead to a discussion about the assumptions that hindered understanding, and which ones led to major breakthroughs in science and technology.

Social-institutional system: Here the focus could be on scientific ethos, professional activities, social certification, competition among scientists (i.e. personalities) in relation to genetic engineering. How does what they did in class resemble what scientists do, how do scientists get better results? Who owns the genetic code? What societal impact does this topic carry? How does the public use this information? What ethical issues confront research in this area? How does knowing the science underlying genetic engineering help students help them sort through media claims? What financial and political issues does this domain entail? Who owns the knowledge that the scientists produce? Is it public or private property? What aspect of this knowledge is proprietary? What role does the legal system play? What role do citizens play? Can the government limit what can be studied? How does all this discussion affect what you do as a student, consumer of goods (e.g. food, medicine) and as future citizens?

In the case of the topic of heredity and diversity at the high school level, we note that more complexity about the social context can be shared at a much more detailed level than was done in the primary science example. But in both examples, all categories and components of the FRA are included. This sort of coverage can be part of a problem-based learning approach, a socio-scientific issues approach, or a more traditional one. The decision of how to contextualize the FRA categories can be tailored to students' interests and abilities and accommodated with the curricular constraints. The final outcome is that by the time the unit is done, students will have learned substantial science and NOS content. They will have covered a broad range of ideas as they reflected on aims and values, practices, methodologies, models, and wide-range of related social issues. Connections can be made between science and engineering practices, genetics, genomics, genetic engineering, legal issues, public interest, privatizing knowledge through patenting laws. The depth and breadth can be pursued in a number of ways: through group projects focusing on one of these facets, through debates in which a jigsaw strategy is used to redistribute expertise across newly formed 'expert' groups, so on and so forth. There is no shortage of ways to organize or sequence the learning of these principles using historical cases or current local and global events.

## 8.2.2 Horizontal Articulation

In the previous section, we illustrated how components of the FRA increase in sophistication as science concepts get more complex in moving from primary to secondary school curriculum. In this section, we outline how the FRA categories can be targeted across science topics taught in the same grade level. A similar process can be followed for outlining how the FRA categories can be connected to the content. This shows how the FRA can help maintain a continuity of coverage of NOS themes throughout the school year. This is a matter of great concern to science educators who have often complained about the typical NOS coverage in an introductory textbook chapter that never gets to be revisited again in successive lessons.

As an example, we start with *Science: Programme of Study for Key Stage 4* (DFE, 2013) from England aimed at the age group 14–16. Table 8.2 illustrates how the FRA categories can be mapped to some example topics (ie. cell biology, Periodic Table and energy). We use subtopics from each main topic to produce example descriptions of FRA categories. Figure 8.4 illustrates how systemic inclusion of the

**Table 8.2** Articulation of FRA categories across science topics in *Science: Programme of Study for Key Stage 4* (DfE, 2013) from England

| Science topic | Cell biology | Periodic table | Energy |
|---|---|---|---|
| Subtopic | *The importance of stem cells in embryonic and adult animals and of meristems in plants* | *Predicting chemical properties, reactivity and type of reaction of elements from their position in the Periodic Table* | *National and global fuel resources, renewable energy sources* |
| Aims and values e.g. Empirical adequacy | Use data on stem cells to determine how they influence embryo development | Use data on the physical and chemical properties of elements to conclude which elements they belong to | Use data on fuel resources and how they provide energy |
| Practices | Discuss similarities and differences between experiments and simulations performed in class and those done in academic or industrial labs | Generate classifications of elements on the basis of their physical and chemical properties; consider how different classification and arrangements of the elements in the Periodic Table illustrate different trends in properties | Generate classifications on the pros and cons of different energy sources and their risks to environment. Generate representations of data produced by scientists noting aspects of practices that explain differences between communities |
| Methods | Compare the different methods scientist use to conduct stem-cell research. Discuss manipulative methods, compared to non-manipulative methods | Conduct experiments to compare chemical reactions of different elements e.g. oxidation and solubility in water | Discussion and comparison of energy production techniques based on a range of energy sources like solar, wind and nuclear energy |
| Knowledge | Consider how stem cell theory fits in with other theories, and how new explanatory models in this area revised our understanding about cell growth and development | Consider the variation between the columns and periods of the Periodic Table and what they indicate about chemical and physical properties of elements | Consider the nature of different sources of energy and compare their efficiency in generating energy |
| Social-institutional e.g. Economic, ethical | Discuss impact of stem cell research on the health sector, medical field, and personal decisions; ethical issues arising from stem cell research; funding issues (public v private) and knowledge ownership | Predict the personal and environmental safety of chemicals and hold institutions responsible for ethical disposal of chemical waste | Consider the political and economic interests governing the use of national and global energy resources, investment in researching green energy sources |
| | | Consider the economic impact of some chemicals (e.g. in food processing industry, in air) on personal and public health | |

8.2 Educational Applications of FRA and GIS

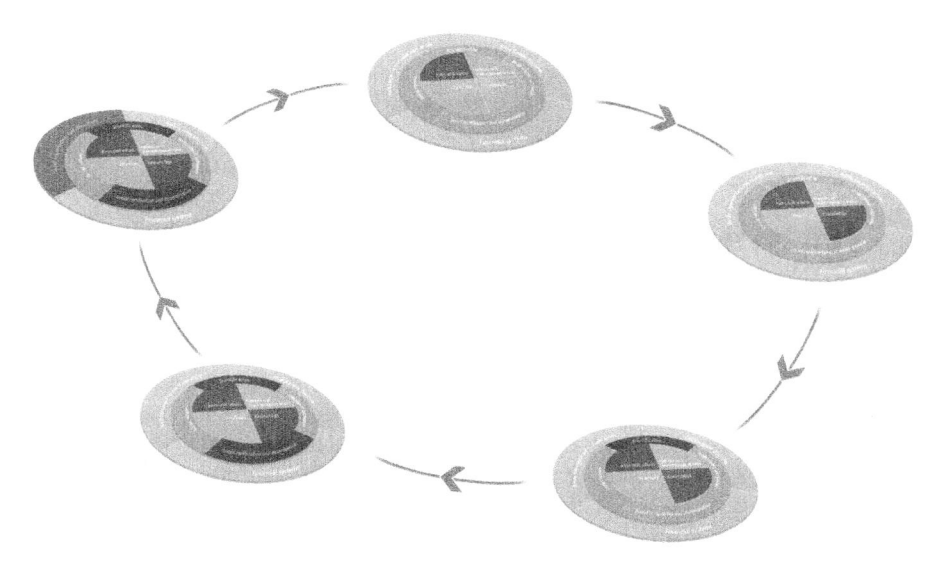

**Fig. 8.4** Rotating emphases for unpacking different categories of the FRA within the same science topic or across topics/domains

FRA categories can be accomplished. As seen here, there is no NOS predetermined content that is simply inserted in each row—but there is a category that makes it possible to bring in relevant NOS 'talking points', specifically tailored to the science content. By the time students finish 12 years of schooling that are focused on the multidimensional approach to NOS, students will have amassed a relatively sophisticated understanding of NOS ideas each of which is contextualized and instantiated in disciplinary or multidisciplinary knowledge. They will develop the ability to transfer this information into novel contexts. They will be prepared, for instance, to ask critical questions about methods, justification for claims, values underlying research projects. They will be able to identify ethical issues, guard against gender inequities in science and other fields, and understand the economics of funding and proprietary rights. As future citizens, whether laborers, lawyers, teachers or scientists, students will be aware of the ways in which scientific knowledge and reasoning can empower them to question claims, verify information, and make informed decisions.

In summary, the ideas implied by the FRA categories are infused into the curriculum while taking into consideration the developmental, cognitive and instructional sequences. The movement from different emphases as the wheel cycles through the curriculum could be conceived (a) within a particular grade level, i.e. where the wheel is introduced and gets unpacked in detail to a select set of subcomponents as the school year unfolds; (b) across grade levels, i.e. where subject matter knowledge gets specialized allowing for the inclusion of additional FRA categories. In either case, the wheel model provides a visual and dynamic representation of the curricular components as well as the instructional processes. Overall, the FRA wheel model and its embedded GIS are comprehensive, dynamic, flexible, and fluid. They can be integrated meaningfully into the science curriculum not only across schooling from kindergarten to high school level but also potentially in postsecondary science education.

## 8.3 FRA, GIS and Curriculum Policy Documents

In discussing the progressive iterations of GIS in schooling, we have advanced the position that a comprehensive account of science (i.e. including the epistemic, cognitive and social aspects) need to be presented to students in a holistic fashion across all grade levels. It was also noted that particular aspects might need to be 'suppressed' while others are emphasized at different grade levels. This position may at times be at odds with some curricular policy arguments that advocate particular aspects of science to be taught at particular grades only. For example, *A Framework for K-12 Science Education* (NRC, 2012) in the USA suggests that economic and other connections to science be explored in Honors or AP courses. Our concern in this regard is that by associating particular aspects of science with particular competency levels, the majority of students who choose not to take advanced courses will not get a good understanding of the social context of science. In practice, we acknowledge that it may not be possible or feasible to address every aspect of the FRA at the same depth at each stage of schooling. However, in principle, aspects from each of the FRA categories must be addressed *systemically* so that students do not end up with fragmented or distorted conceptualization of NOS. This is where vertical and horizontal alignment within and across the science curriculum would help maintain coherence.

The alignment of the curriculum with learning and assessment goals is a necessary undertaking. In using the FRA for curriculum planning, it is important to match curricular goals with innovative instructional interventions and assessment forms. Douglas Allchin who has argued for the teaching and learning of the nature of (whole) science (Allchin, 2011) also highlights the significance of designing assessments that are consistent with revised instructional goals. Before turning to the issue of assessment in more detail, example standards from *NGSS* (NGSS Lead States, 2013a) are used to illustrate how the FRA can serve as a framework to investigate science standards.

### 8.3.1 Example 1: HS-LS3 Heredity: Inheritance and Variation of Traits

In this example, heredity-related standards are examined against the main categories of the FRA to model how NOS ideas can be integrated into these content standards. The "Heredity: Inheritance and Variation of Traits" standard (NGSS Lead States, 2013a, p. 89) includes three main learning outcomes:

## 8.3 FRA, GIS and Curriculum Policy Documents

HS-LS3-1. Ask questions to clarify relationships about the role of DNA and chromosomes in coding the instructions for characteristic traits passed from parents to offspring.

HS-LS3-2. Make and defend a claim based on evidence that inheritable genetic variations may result from: (1) new genetic combinations through meiosis, (2) viable errors occurring during replication, and/or (3) mutations caused by environmental factors.

HS-LS3-3. Apply concepts of statistics and probability to explain the variation and distribution of expressed traits in a population.

The emphasis in this standard is on asking questions for the purpose of clarifying relationships. The standard focuses on developing evidence-based arguments and the engagement in mathematical thinking through the use of statistics and probability concepts. Students are expected to engage in scientific practices along the lines defined in the vision document (NRC, 2012): ask questions, make and defend claims, and use mathematical thinking. However, understanding scientific knowledge and practices does not guarantee understanding the nature of science. The "connections to nature of science" included in the interpretive section below the standards specify two ideas under "science as a humans endeavor" theme:

- Technological advances have influenced the progress of science and science has influenced advances in technology. (HS-LS3-3)
- Science and engineering are influenced by society and society is influenced by science and engineering. (HS-LS3-3)

The connections to nature of science called for here are too broad to help teachers determine the relationships that should be made between science and technology, and science and society. It will be left to the imagination of curriculum developers to tie these two connections to nature of science to the third standard. The primary learning outcomes expressed in the standards do not *explicitly* include science as a human endeavor (an NOS Category in Appendix H), nor require an understanding of NOS dimensions discussed by the FRA: scientific aims and values, practices, methods, knowledge and science as social systems in science including science as a social enterprise. In other words, it is possible for students to understand the explicitly stated learning outcomes and bypass the meta-level connections with science as a cognitive-epistemic system or science as a social-institutional system. If the learning goals miss a holistic account of science, so will the instruction. If particular aspects of science are not prioritized nor signaled as important or relevant in applying the standards to the curriculum, they will not be assessed.

The HS-LS-3 Standard misses a golden opportunity to address scientific ethical, social, and economic arguments pertaining to the developing technologies, funding and ethical issues in the context of genetic mutations and genetically modified organisms. Genetic modification technologies bring up significant issues that affect personal and societal decisions about safety, cost, social and environmental impacts. They also provide a meaningful context for discussing the role of patents in limiting access not only to the products of those technologies but also to scientific knowledge itself. The standard does not refer to genetic modification technologies of cloning, gene therapy, genetic engineering and selective breeding.

### 8.3.2 Example 2: From Molecules to Organisms – Structures and Processes

The next example, based on high school life science standards in the NGSS (NGSS Lead States, 2013a), is used to illustrate how individual standards pertaining to this topic can be supplemented with NOS content. Because a NOS meta-cognitive structure is lacking in the current standards, we exemplify in Table 8.3 how the FRA can be used as a tool to guide the selection of appropriate NOS content that complement and enrich the life science standards depicted here. Expressing the infusion statements in terms of learning outcomes provides clarity for instructional and assessment purposes.

The GIS serve as heuristics or memory aids in searching for NOS content. The example in Table 8.3 illustrates using the FRA as an analytical tool that can be further refined to provide a fine-grained analysis. The comments in the third column of the Table show the results of conducting a systematic content analysis on the target standards conducted to identify NOS connections, strengthen those that are weak, and address those that are absent. For a systematic evaluation or curriculum development purposes, it is useful to study standards pertaining to a given grade band (e.g. K-2, 3–5) to identify missing components and develop amendments. In summary, the FRA and the associated GIS can be 'tweaked' for use as meta-cognitive tools to analyze, evaluate, or reflect on curriculum materials or curriculum policy documents in order to identify if and how different aspects of NOS are being addressed, and help develop a coherent plan for addressing pertinent but missing FRA categories in science curriculum and instruction.

## 8.4 Potential limitations of the FRA and GIS

The extended FRA categories and the related GIS are intended to provide a multifaceted approach to addressing a range of ideas that impact NOS and its teaching. At the same time, it should be noted that the FRA has a number of potential limitations. These limitations are summarized in relation to approximation, ontology, metaphysics, openness, application and perception of competing goals. In the following paragraphs, each of these issues is discussed to alert the reader about where further work is needed in developing the strengths of FRA.

*Approximation:* The FRA wheel can be used as an instructional model that approximates the components of a complex domain (i.e. science and nature of science). It brings coherence to science content by uniting it around salient FRA categories. The cognitive and developmental aspects of the various categories of science represented in GIS need to be researched by using empirical evidence on teaching and learning such that theoretical rationales for using FRA/GIS in science education are complemented with empirical ones.

8.4 Potential limitations of the FRA and GIS    177

Table 8.3 HS-LS1 from molecules to organisms: structures and processes in the NGSS (NGSS Lead States, 2013a)

| GIS | Standard HS-LS1 | Comments to support FRA infusion |
|---|---|---|
| Aims and values | Not addressed | Several aims and values can be addressed in this context. Focusing on criteria that scientists use to evaluate the strength of evidence can be one aim that could be easily included |
| | | *To infuse NOS, add:* |
| | | Generate criteria to evaluate and reflect on the quality of data, and if necessary revise criteria for data evaluation |
| Practices | HS-LS1-1. *Construct an explanation based on evidence for how the structure of DNA determines the structure of proteins which carry out the essential functions of life through systems of specialized cells* | These learning outcomes involve the design and performance of an investigation, which are components of scientific activities but they fall short of adding the reflective component necessary for discussing the nature of science |
| | HS-LS1-2. *Develop and use a model to illustrate the hierarchical organization of interacting systems that provide specific functions within multicellular organisms* | *To infuse NOS, add:* |
| | HS-LS1-3. *Plan and conduct an investigation to provide evidence that feedback mechanisms maintain homeostasis* | Consider the nature of representations, particularly in relation to models and modeling. Produce and evaluate different representations. Some of the evaluation criteria generated in Chap. 6 (e.g. projectability, approximation) can be relevant in establishing how models are articulated and evaluated |
| | HS-LS1-4. *Use a model to illustrate the role of cellular division (mitosis) and differentiation in producing and maintaining complex organisms* | |
| | HS-LS1-7. *Use a model to illustrate that cellular respiration is a chemical process whereby the bonds of food molecules and oxygen molecules are broken and the bonds in new compounds are formed resulting in a net transfer of energy* | |

(continued)

**Table 8.3** (continued)

| GIS | Standard HS-LS1 | Comments to support FRA infusion |
|---|---|---|
| Methods | HS-LS1-3. Plan and conduct an investigation to provide evidence that feedback mechanisms maintain homeostasis | This learning outcome can be supported from an NOS perspective by discussing a *range of relevant methods for designing appropriate investigations* |
| | | *To infuse NOS, add:* |
| | | Evaluate the pros and cons of different investigations. Determine what methods are likely to produce stronger evidence. Compare to methods and tools to those of scientists |
| Knowledge | HS-LS1-5. Use a model to illustrate how photosynthesis transforms light energy into stored chemical energy | This learning outcome supports student use of existing models, or constructing and revising model. It misses the metacognitive components that link their work with models to the work of scientists |
| | HS-LS1-6. Construct and revise an explanation based on evidence for how carbon, hydrogen, and oxygen from sugar molecules may combine with other elements to form amino acids and/or other large carbon-based molecules | *To infuse NOS, add:* |
| | HS-LS1-7. Use a model to illustrate that cellular respiration is a chemical process whereby the bonds of food molecules and oxygen molecules are broken and the bonds in new compounds are formed resulting in a net transfer of energy | Reflect on the nature of models. Compare students' models to current or historical models. Discuss the bases on which these models can be, or have been, evaluated |

8.4 Potential limitations of the FRA and GIS 179

| Social-institutional system | | |
|---|---|---|
| *Professional activities* | Not addressed | *To infuse NOS, add:* Examine scientific literature and prepare presentations to communicate findings or defend recommended models |
| *Scientific ethos* | Not addressed | *To infuse NOS, add:* Discuss the rights of human subjects when obtaining genetic material |
| *Social certification and dissemination of scientific knowledge* | Not addressed | *To infuse NOS, add:* Share and debate written/oral presentations. Compare process to communities of scientists |
| *Social values* | Not addressed | *To infuse NOS, add:* Explore ways in which social utility as a value has contributed to this line of work. |
| *Social organization and interaction* | Not addressed | *To infuse NOS, add:* Discuss the variety of scientific networks working on DNA research |
| *Political power structures* | Not addressed | *To infuse NOS, add:* Discuss relevant cases that illustrate gender bias and political influences on genetics research |
| *Financial systems* | Not addressed | *To infuse NOS, add:* Discuss cases that illustrate the financial dimensions of genetics research |

*Ontology:* A potential limitation is that the GIS are based on cognitive, epistemic and social- institutional dimensions of science, and does not focus on ontological assertions. However, in the Benzene Ring heuristic, we have referred to a real world that scientific practices deal with. Aspects of GIS can be used to raise discussions about ontological assumptions.

*Metaphysics:* One important aspect of nature of science pertains to its metaphysical assumptions which were not explicitly and directly addressed in this book. Three metaphysical assumptions are highlighted by Dilworth (2007): (a) the principle of uniformity of nature, (b) the principle of substance, and (c) the principle of causality. FRA does not explicitly deal with Dilworth's metaphysical assumptions. The NGSS include a statement that falls under Dilworth's first principle. It states that "scientific knowledge assumes an order and consistency in natural systems" (NGSS Lead States, 2013b, p. 6). We believe that science education should instill in learners some awareness about important assumptions that form the foundations of science.

*Openness:* The FRA has a generative nature that we consider to be one of its strengths as well as its limitation. In the same way that this generative nature can inspire creative means for enriching teaching and learning, it is possible that some future depictions extend the features of science to dimensions that we do not yet anticipate nor endorse. Hence FRA is inherently prone to exploitation and distortion. We suggest that however the FRA categories are extended, the best policy is to use an evidence-based approach to their articulation akin to the way that we have drawn from the research literature on philosophy of science.

*Application:* The variety of FRA categories and the related GIS serve as metacognitive tools that are dependent on a relatively good understanding of the cultural studies of science. While the GIS provide a reminder of which components to include within each FRA category, the application of the images demands careful research and selection of supplementary materials (such as historical episodes and methodological case studies) that are inherently specific to the science content.

*Perception of competing goals:* The FRA must not take on a life of their own. FRA categories do not compete with but rather serve broader science education goals, such as the holistic representation of science in school science. However the FRA runs the risk of being misperceived as placing unreasonable demands on the curriculum. The optimal use of FRA and the related GIS is heavily dependent on integration with core science concepts. These concepts become the context where reflective consideration of scientific values, practices, methods, knowledge and social processes can take place.

The implementation of the FRA can be facilitated by knowing what GIS can and cannot do, and considering the ways in which the strengths can be optimize and the limitations minimized. Despite the mentioned potential limitations, the FRA provides a fruitful reconceptualization of NOS in science education. It also provides innovative avenues for future research. For example, investigating the extent to which students' understanding of NOS might improve given a holistic and visual account of science is a line of work that is at the heart of the empirical validation of the proposed FRA framework.

## 8.5 Recommendations

The previous sections illustrated multiple ways in which the FRA and the GIS can be used as tools to articulate NOS ideas vertically and horizontally across the curriculum, and to analyze curriculum standards documents. In this section, some recommendations for teaching, teacher education, curriculum and assessment are made in order to support the implementation of FRA categories in school science. We are mindful of the fact that the reconceptualized version of NOS is a theoretical account and hence, the recommendations are meant to be guidelines that can help inform researchers who are interested in pursuing future empirical studies.

### 8.5.1 Teaching

The FRA and the related GIS are likely to be effectively taught when teachers couple them with evidence-based science learning strategies. Some strategies that have been extensively researched include the use of practical inquiries, group discussions and presentations, role play, questioning, differentiation and peer assessment (e.g. Abell & Lederman, 2007; Gabel, 1993; Palincsar, Magnusson, Collins, & Cutter, 2001). Model-based inquiries immerse students in investigations where they collect, interpret and present data to generate scientific explanations, models and arguments. Group discussions and presentations engage learners in the social and cultural practices of science through communication, dialogue and public display of ideas based on evidence. Role play enable students to evaluate different points of view including a range of explanations for a particular phenomenon; it engages learners in the generation and application of criteria for discriminating scientific ideas from other ways of knowing. Differentiation provides the opportunity to tailor the science content to the needs and abilities of individual students. Peer assessment promotes student voice in the classroom and creates a context for learning among peers. Such strategies represent a sample of teaching approaches that promote active communities of learning and personal engagement with science. Furthermore, such strategies model ways of acting, thinking and communicating that form the fabric of the culture of science as a discipline. For example, scientists themselves argue about different hypotheses, theories and models; science cultures tend to have a range of expertise where problems to be investigated are differentiated according to background and interests; professional peer review systems validate and justify the dissemination of scientific knowledge.

### 8.5.2 Teacher Education

There is a substantial body of literature on teachers' continuous professional development (CPD). The effective uptake of the FRA by teachers will rely on the incorporation of evidence on CPD. Within proposed and researched CPD models, it is widely accepted that learning to teach is not a linear process and that educational change is not a "natural consequence of receiving well-written and comprehensive instructional materials" (Hoban, 2002, p. 13). For teachers' learning to be effective, a more complex view of professional development is required, incorporating professional learning systems. It is widely documented that educational change is complex and takes time (Fullan, 2001), and fundamental and substantial changes could not be achieved within a short period of time (e.g. Erduran & Dagher, 2007). Furthermore, across the world, in the current context of accountability and high stakes assessment, teachers operate within curricular constraints that may be perceived to be incompatible with innovative approaches to teaching and learning.

Supovitz and Turner (2000) argue that high-quality professional development (a) immerses participants in inquiry, questioning and experimentation; (b) is intensive and sustained; (c) engages teachers in concrete teaching tasks and is based on teachers' experiences with students; (d) focuses on subject-matter knowledge and deepen teachers' content skills; (e) is grounded in a common set of professional development standards and show teachers how to connect their work to specific standards; and (f) is connected to other aspects of school change. Effective teacher education however often requires teachers to engage in practices that may not be supported by institutional expectations, for example sharing of teaching resources versus maintaining privacy about them (Spillane, 1999). Apart from teachers experimenting with new strategies, teachers' reflections on their practices are essential part of their learning. However it is difficult to anticipate the extent to which any new professional development initiative would facilitate the process of reflection-in-action, or reframing (Schön, 1987), that results in constructing new pedagogical understanding of NOS.

Nevertheless articulation of teachers' knowledge about FRA will need to be consistent with successful models in teacher education research. For example, Shulman and his colleagues' conceptualization of teachers' subject-matter knowledge in terms of "content knowledge", "pedagogical content knowledge" (PCK), and "curricular knowledge" are significant constructs to apply to the FRA because such application may illustrate what teachers will need to know in order to teach NOS based on a FRA. According to Shulman (1986), "content knowledge" refers to "the amount and organization of knowledge per se in the mind of the teacher", including knowledge of the "substantive structure" and "syntactic structure" of the academic discipline — two terms borrowed from Joseph Schwab (1964). The syntactic structure concept, for instance, can be embellished with a broader framing provided by the FRA. For example, the issue of growth of scientific knowledge as highlighted in Chap. 6 can pinpoint the ways in which knowledge construction mechanisms operate in science. Also named "subject-matter knowledge for teaching", "content knowledge" was subsequently elaborated upon by Grossman, Wilson, and Shulman (1989) as consisting of the following four components: (a) content knowledge—the "stuff" of a discipline; (b) substantive knowledge—knowledge of the explanatory framework or paradigms of a discipline; (c) syntactic knowledge—knowledge of the ways in which new knowledge is generated in a discipline; and (d) beliefs about the subject matter—feelings and orientations toward the subject matter. All of these components of teacher knowledge are directly relevant to applying the FRA categories to teacher professional development. For example, extending the teachers' knowledge base of the social and institutional aspects of science may provide a fruitful territory for teacher educators to consider in relation to how teachers can integrate such aspects into their existing pedagogical repertoires.

### 8.5.3 *Curriculum and Assessment*

Throughout this chapter and the previous chapters, we have repeatedly drawn on example curriculum standards to illustrate the relevance and utility of the FRA for curriculum planning and design. Hence the recommendations in relation to the curriculum are situated in each aspect of FRA along with the suggestions on how FRA can help improve the content of the curriculum. As a summary, a FRA can:

1. provide models in developing and implementing teaching units and lesson plans;
2. promote discussion of relevant epistemic, cognitive and social-institutional issues in relation to curriculum content;
3. establish focus for the exploration of historical or contemporary science cases (such as those described by Allchin, 2013), or researching recent news reports, where the cases are relevant thematically and developmentally to the target audience;
4. serve as a point of reference for exploring the content of science topics from as many angles (e.g. epistemic, cognitive, social, cultural, financial) as possible.

The GIS produced in each chapter can be used as starting points for developing more specific assessment tools for use in teacher preparation programs as well as K-12 classrooms. For example, indicators of understanding the aims and values of science can be generated. An example of scientific aims and values reviewed in Chap. 3 is "empirical adequacy". Theoretical accounts of such scientific aims and values can be scrutinized relative to the research evidence on how children use data and evidence in supporting their claims derived from empirical investigations (e.g. Kuhn, 1991). Overall, developing a functional use of the FRA is contingent on establishing coherence between the curriculum and assessment goals.

## 8.6 Contributions to Research and Practice in Science Education

The FRA and GIS have the potential to contribute to various aspects of research in science education. As tools for conceptual analysis, they can be used to examine research on nature of science or research in science education in general. They can help identify trends in the research literature. For example, GIS can help query to what extent the economical aspects of science have been a research focus in science education. In a similar vein, GIS can help further articulate existing bodies of research. If studies historically fell into one or two categories of the FRA, the GIS provides a chance to reflect on where else the work could go next. For example, work on argumentation in science education typically covered argumentation as a particular instance of scientific practices and scientific knowledge (e.g. Erduran, 2007; Erduran, Simon, & Osborne, 2004). GIS can help identify missing NOS aspects, for example the impact of organizational structures in the validation of scientific arguments and implications for the design of learning environments. As analytical heuristics, they can help identify various trends and emphases as well as missing aspects of NOS in science education research and policy.

The expanded FRA and the GIS articulate and reconcile the tensions between a set of nature of science ideas that are rooted in general principles that cut across the sciences and nature of science ideas that are rooted in specific science domains. What are the GIS? Are they generic or domain-specific? We contend that they are both. They are generic in terms of the broad category, such as scientific practices. However, the category is vacuous without the content-specific details. The teaching content of an FRA category is bound by reflective thinking of a specific domain. This reflective thinking emerges from insights gained from philosophical, historical, social and cultural studies of science. It is useless to talk about generic practices, generic methods without pointing to specific practices and specific methods from which these generic ones were derived. On its own, we can take any FRA category, for example methods and methodological rules, and discuss methodological possibilities, but that is not the point we are trying to make in our treatise of FRA in science education. By the same token, it can be argued that methods are always taught in connection with science. In the absence of a reflective component anchored in a particular science domain, the relevance of the diversity of methods could be easily missed.

To continue with the example of the methods category, the purpose for discussing these methods is to communicate how different fields of study lean on a variety of *specific* methods. Even though emphasis on the various methods differs across science domains (e.g. the role of experimentation in astronomy versus chemistry), noting these differences in the context of progressing through different domains across the science curriculum in a school year or an entire educational career provides opportunities for building a profound understanding of the range of methods scientists use to generate trustworthy findings. Domain-specificity can also contribute to complex theoretical narratives leading to deeper understanding of not only the methods themselves but also the nature of the knowledge that is generated through the deployment of such methods. This is very different from knowing that scientific methods are diverse at a superficial level. Our approach forces understanding the roots of methodological diversity, why is it useful and what it achieves. In this sense, any serious characterization of scientific methods in general cannot escape the domain-specificity of scientific methods. The following quote elegantly addresses a parallel relationship between science domains as parts and the whole of science:

Parts and wholes evolve in consequence of their relationship, and the relationship itself evolves. These are the properties of things that we call dialectical: that one thing cannot exist without the other, that one acquires its properties from its relation to the other, that the properties of both evolve as a consequence of their interpenetration. (Levins & Lewontin, 1985, p. 3)

A related but different issue concerns the potential of the FRA to facilitate meta-cognitive awareness of the domain-specific aspects of science. While the FRA is based on an approach that approximates similarities between the various branches of science, it also organizes thinking around the kinds of differences that might exist in branches of science. For example, as discussed in Chap. 6, the way in which chemists and physicists understand 'laws' may have some differences. While various branches of science might have laws as part of the scientific knowledge repertoire, the disciplinary variations can be highlighted.

## 8.7 Conclusions

The GIS and the FRA on which they are based, provide science educators, specifically teachers and researchers, with heuristic tools for situating scientific values, knowledge, methods, practices and social-institutional systems in ways that can potentially motivate students. These tools promote understanding of science as the interplay of a cognitive-epistemic-social-institutional dynamic that is constantly developing and evolving. Like "scientists [who] produce new knowledge in many domains through generating and analyzing the content of images" (Prain & Tytler, 2013, p. 1), as educators we sought to generate images about science for the purpose of opening up conversations on practical pathways for enriching science teaching and learning. Even though a range of examples were presented in each chapter, it is important to envision the totality of these images in use. For instance, it is vital for educators to consider the content they impart, and how the GIS might be infused within a unit of study, across units of study in a school year, and across an entire K-12 education. The images are iconic meta-cognitive tools that can help teachers and learners consider the nature of scientific aims and values, the nature of data, evidence, arguments and models, and the nature of social values as they operate within the scientific community and the larger society. We hope readers will be inspired to use these tools to support teaching and research agenda in K-12 schools and teacher education settings.

The book is broadly related to the science education research literature on NOS. However, within the historical progression of NOS (e.g. Abd-el-Khalick & Lederman, 2000; Khishfe & Abd-el-Khalick, 2002; Lederman, 1992, 2007; Schwartz, Lederman, & Crawford, 2004) research has been limited in providing a holistic and visual account of nature of science. The holistic aspect relates to the coordination of the cognitive, epistemic and social-institutional dimensions of science while the visual aspect refers to the transformation of such dimensions to visual representations that can be effectively used in application to science education. The GIS provide some practical heuristics with which researchers, curriculum reformers and science teachers can articulate the complexity of NOS in science education.

Our articulation of the FRA categories is related to disparate areas of research in science education, such as studies on socio-scientific issues (e.g. Zeidler, Sadler, Simmons, & Howes, 2005), inquiry-based science teaching and learning (e.g. Duschl & Grandy, 2008; Welch, Klopfer, Aikenhead & Robinson, 1981), metacognition (e.g. Zohar & Dori, 2012), argumentation (e.g. Erduran & Jimenez-Aleixandre, 2008), critical thinking (e.g. Bailin, 2002; Zoller, 1999), history and philosophy of science (e.g. Duschl, 1990; Matthews, 1994), and learning progressions (e.g. Duschl, Maeng, & Sezen, 2011). However our reconceptualization of NOS goes beyond the particular research traditions listed here. In articulating perspectives from philosophy of science, we have (a) appealed to a coherent theoretical rationale on NOS proposed by philosophers of science, (b) developed this theoretical framework extensively, and (c) anchored the extended framework in science curriculum, teaching and learning.

In exploring the interplay between philosophy of science and science education, we have been mindful of the curricular, research and policy contexts of science education, thus drawing on some evidence from these accounts as well selecting perspectives that can have utility and appeal in science education. Ultimately, however, our approach is motivated by a belief that the FRA and GIS will empower learners to engage in science and use their understanding effectively to improve the quality of their lives and the well-being of their communities. Our hope is that the perspectives developed in this book will foster discussion and research for the improvement of science teaching and learning for all students.

# References

Abd-El-Khalick, F., & Lederman, N. G. (2000). Improving science teachers' conceptions of nature of science: A critical review of the literature. *International Journal of Science Education, 22*(7), 665–701.

Abell, S., & Lederman, N. (Eds.). (2007). *Handbook of research in science education, Part 3*. New York: Routledge, Taylor & Francis.

Allchin, D. (2011). Evaluating knowledge of the nature of (whole) science. *Science Education, 95*(3), 518–542.

Allchin, D. (2013). *Teaching the nature of science: Perspectives and resources*. St. Paul, MN: SHiPs.

Bailin, S. (2002). Critical thinking and science education. *Science & Education, 11*, 361–375.

Chalfour, I., & Worth, K. (2006). *Science in kindergarten*. Retrieved July 31, 2013 from http://www.rbaeyc.org/resources/Science_Article.pdf

Department for Education. (2013). Science: Programme of Study for Key Stage 4 (February 2013). Retrieved on August 4, 2013, from http://media.education.gov.uk/assets/files/pdf/s/science%20-%20key%20stage%204%2004-02-13.pdf

Dilworth, G. (2007). *The metaphysics of science: An account of modern science in terms of principles, laws and theories*. Dordrecht, The Netherlands: Springer.

# References

Duschl, R. A. (1990). *Restructuring science education: The importance of theories and their development.* New York: Teachers College Press.

Duschl, R., & Grandy, R. (Eds.). (2008). *Teaching scientific inquiry: Recommendations for research and implementation.* Rotterdam, The Netherlands: Sense Publishers.

Duschl, R., Maeng, S., & Sezen, A. (2011). Learning progressions and teaching sequences: A review and analysis. *Studies in Science Education, 47*(2), 119–177.

Erduran, S. (2007). Breaking the law: Promoting domain-specificity in science education in the context of arguing about the Periodic Law in chemistry. *Foundations of Chemistry, 9*(3), 247–263.

Erduran, S., & Dagher, Z. (2007). Exemplary teaching of argumentation: A case study of two middle school science teachers. In R. Pinto & D. Couso (Eds.), *Contributions from science education research* (pp. 403–415). Dordrecht, The Netherlands: Springer.

Erduran, S., & Jimenez-Aleixandre, M. P. (Eds.). (2008). *Argumentation in science education: Perspectives from classroom-based research.* Dordrecht, The Netherlands: Springer.

Erduran, S., Simon, S., & Osborne, J. (2004). TAPping into argumentation: Developments in the application of Toulmin's argument pattern for studying science discourse. *Science Education, 88*, 915–933.

Fullan, M. (2001). *The new meaning of educational change* (3rd ed.). London: Routledge-Falmer.

Gabel, D. L. (Ed.). (1993). *Handbook of research in science teaching and learning project.* New York: Macmillan.

Gilbert, J. (Ed.). (2005). *Visualisation in science education.* Dordrecht, The Netherlands: Springer.

Gilbert, J. K., Reiner, M., & Nakhleh, M. (Eds.). (2008). *Visualization: Theory and practice in science education.* Dordrecht, The Netherlands: Springer.

Grossman, P. L., Wilson, S. M., & Shulman, L. S. (1989). Teachers of substance: Subject matter knowledge for teaching. In M. C. Reynolds (Ed.), *Knowledge base for the beginning teacher* (pp. 23–36). New York: Pergamon.

Hoban, G. F. (2002). *Teacher learning for educational change.* Buckingham, UK: Open University Press.

Irzik, G., & Nola, R. (2014). New directions for nature of science research. In M. Matthews (Ed.), *International handbook of research in history, philosophy and science teaching* (pp. 999–1021). Dordrecht, The Netherlands: Springer.

Johnson-Laird, P. N. (1998). Imagery, visualization, and thinking. In J. Hochberg (Ed.), *Perception and cognition at century's end* (pp. 441–467). San Diego, CA: Academic.

Khishfe, R., & Abd-El-Khalick, F. (2002). Influence of explicit and reflective versus implicit inquiry-oriented instruction on sixth graders' views of nature of science. *Journal of Research in Science Teaching, 39*(7), 551–578.

Kridel, C. (Ed.). (2010). *Encyclopedia of curriculum studies* (Vol. 1). Thousand Oaks, CA: Sage. doi: http://dx.doi.org/10.4135/9781412958806, doi: 10.4135/9781412958806#blank

Kuhn, D. (1991). *The skills of argument.* New York: Cambridge University Press.

Lederman, N. G. (1992). Students' and teachers' conceptions of the nature of science: A review of the research. *Journal of Research in Science Teaching, 29*(4), 331–359.

Lederman, N. G. (2007). Nature of science: Past, present, future. In S. Abell & N. Lederman (Eds.), *Handbook of research on science education* (pp. 831–879). Mahwah, NJ: Lawrence Erlbaum.

Levins, R., & Lewontin, R. (1985). *The dialectical biologist.* Boston, MA: Harvard University Press.

Matthews, M. (1994). *Science teaching: The role of history and philosophy of science.* New York: Routledge.

National Research Council. (2012). *A framework for k-12 science education.* Washington, DC: National Academies Press.

NGSS Lead States. (2013a). *Next generation science standards: For states by states.* Retrieved from http://www.nextgenscience.org/next-generation-science-standards

NGSS Lead States. (2013b). *Next generation science standards: For states by states*. Appendix H. Retrieved from http://www.nextgenscience.org/next-generation-science-standards

Palincsar, A. S., Magnusson, S. J., Collins, K. M., & Cutter, J. (2001). Making science accessible to all: Results of a design experiment in inclusive classrooms. *Learning Disability Quarterly, 24*, 15–32. matth.

Phillips, L. M., Norris, S. P., & Macnab, J. S. (2010). *Visualization in mathematics, reading and science education*. Dordrecht, The Netherlands: Springer.

Prain, V., & Tytler, R. (2013). Representing and learning in science. In R. Tytler, V. Prain, P. Hubber, & B. Waldrip (Eds.), *Constructing representations to learn in science* (pp. 1–14). Rotterdam, The Netherlands: Sense Publishers.

Schön, D. (1987). *Educating the reflective practitioner: Toward a new design for teaching and learning in the professions*. San Francisco: Jossey-Bass.

Schwab, J. J. (1964). The structure of the disciplines: Meaning and significance. In G. W. Ford & L. Pugno (Eds.), *The structure of knowledge and the curriculum* (pp. 6–30). Chicago: Rand McNally.

Schwartz, R. S., Lederman, N. G., & Crawford, B. A. (2004). Developing views of nature of science in an authentic context: An explicit approach to bridging the gap between nature of science and scientific inquiry. *Science Education, 88*(4), 610–645.

Shulman, L. S. (1986). Those who understand: Knowledge growth in teaching. *Educational Researcher, 15*(2), 4–14.

Spillane, J. S. (1999). External reform initiatives and teachers' efforts to reconstruct their practice: The mediating role of teachers' zones of enactment. *Journal of Curriculum Studies, 31*(2), 143–175.

Supovitz, J. A., & Turner, H. M. (2000). The effects of professional development on science teaching practices and classroom culture. *Journal of Research in Science Teaching, 37*(9), 963–980.

Welch, W. W., Klopfer, L. E., Aikenhead, G. S., & Robinson, J. T. (1981). The role of inquiry in science education: Analysis and recommendations. *Science Education, 65*(1), 33–50.

Wu, H. K., Krajcik, J. S., & Soloway, E. (2001). Promoting understanding of chemical representations: Students' use of a visualization tool in the classroom. *Journal of Research in Science Teaching, 38*(7), 821–842.

Zeidler, D., Sadler, T., Simmons, M., & Howes, E. V. (2005). Beyond STS: A research-based framework on socioscientific issues education. *Science Education, 89*, 357–377.

Zohar, A., & Dori, Y. J. (Eds.). (2012). *Metacognition in science education: Trends in current research*. Dordrecht, The Netherlands: Springer.

Zoller, U. (1999). Scaling up of higher-order cognitive skills-oriented college chemistry teaching. *Journal of Research in Science Teaching, 36*(5), 583–596.

# Authors Biographies

**Sibel Erduran** is Professor of STEM Education at University of Limerick, Ireland where she is the Director of the National Centre for Excellence in Mathematics and Science Teaching and Learning. She is an Editor for *International Journal of Science Education* and Section Editor for *Science Education*. She has held Visiting Professorships at Kristianstad University, Sweden and Bogazici University, Turkey. She has worked at University of Bristol and King's College London, United Kingdom. She is a member of the Royal Irish Academy Social Sciences Committee and the IHPST Council, and served as the NARST International Coordinator. Her higher education was completed in the USA at Vanderbilt (Ph.D.), Cornell (M.Sc.) and Northwestern (B.Sc.) Universities. She has been a chemistry teacher in a high school in northern Cyprus. Her research interests focus on the applications in science education of interdisciplinary perspectives on science, particularly the epistemic practices of science.

**Zoubeida R. Dagher** is Professor of Science Education at the School of Education and a Faculty Fellow at the Center for Science, Ethics & Public Policy at the University of Delaware. She is currently President-Elect of the International History, Philosophy and Science [IHPST] and has served as an elected member to the Board of Directors of NARST. She has also served as member of editorial review boards in lead science education journals. She has been a Visiting Scholar at Curtin University of Technology, Perth, Australia, the Lebanese University and the American University of Beirut in Lebanon. She received her Ph.D. in science education from the University of Iowa, USA, and her Masters and Bachelor degrees from the American University of Beirut and the Lebanese American University, respectively. Her research focuses on the nature of science and representations of scientific epistemology in science curricula.